Microbiological Contamination Control in Pharmaceutical Clean Rooms

Microbiological Contamination Control in Pharmaceutical Clean Rooms

Edited by
Nigel Halls

CRC PRESS

Boca Raton London New York Washington, D.C.

Sue Horwood Publishing

Library of Congress Cataloging-in-Publication Data

Halls, Nigel A.
Microbiological contamination control in pharmaceutical clean rooms / edited by Nigel Halls.

p. cm.
Includes bibliographical references and index.
ISBN 0-8493-2300-2
1. Drugs—Microbiology. 2. Drugs—Sanitation. 3. Clean rooms. I. Halls, Nigel A., 1945-

RS199.M53 M535 2004
615/.19 22—dc22 2004047804

International Standard Book Number 0-8493-2300-2
Library of Congress Card Number 2004047804
Printed in the United States of America 1 2 3 4 5 6 7 8 9 0
Printed on acid-free paper

Contents

Preface

I am of the opinion that few or any people other than reviewers ever read the prefaces to books. Nonetheless my publisher has asked me to write one!

Contamination control in pharmaceutical clean rooms is a curious subject in the sense that the way in which it is achieved in any particular application is a jumble of science and engineering, of knowledge of what has worked well or badly in the past, of the technology available at the time the clean-room was built, and of subsequent technological developments. Surrounding it all is a blanket of regulations, some of which were written years ago and stood the test of time, some of which are currently evolving through drafts published for review, and some of which are appearing as if by spontaneous generation as inspectors and auditors feel obliged to react to situations they fear to be posing unacceptable risks (real or imagined).

Successful contamination control in pharmaceutical clean rooms calls for a multidisciplinary approach. Within an operational facility the microbiologists have their part in contamination control and monitoring, and the engineers theirs; so too have the production personnel, the quality, validation, logistics, technology transfer and compliance specialists. They have to communicate well and understand each other's difficulties, they have to share knowledge, and they have to accept that responsibilities often overlap. They should appreciate that the greatest risks to contamination control most often occur at interfaces, not just at physical interfaces between areas designated for activities of differing vulnerabilities in the factory, but also at organisational and cultural interfaces between departments, and around topics where personnel with differing educational and vocational backgrounds are obliged to interact.

This book does not and was never intended to comprehensively address all aspects of contamination control in pharmaceutical clean rooms. It is a collection of monographs written by authors who want to share their knowledge, their experience and their opinions on topics that I believe are of importance and should be of interest to all those who are involved in contamination control in pharmaceutical clean rooms.

I have written the first several chapters. When I began to write these chapters, I set out to get back to the basics of contamination control and relate them to practical situations pertinent to a general readership. Now, with 20:20 hindsight, I fear this is an impossible task because in the end no two pharmaceutical clean rooms are the same and what I have grown to believe is the "norm" is, from my experience, actually completely different from what others have come to believe to be the "norm," but frequently via routes of different experience. I also have developed a greater sympathy for the writers of regulatory guidance than I might have had in the past. It must be a very difficult job to create generic rules for an essentially eclectic industry.

I have tried to indicate what is scientifically factual, what is opinion born of personal experience and what is regulation, but I should emphasise that it is well advised in a highly regulated commercial arena to take a conservative tack when interpreting regulations. For instance, a delay of six months in gaining regulatory approval for new drug product should in my opinion be perceived in terms of the profit to be made from that product at its peak volumes under patent protection; let us imagine perhaps $1,000,000. Who would gamble that sort of sum in relation to saving $10,000 dollars on a piece of monitoring equipment, which if purchased, would guarantee compliance with the most conservative interpretation of current regulation?

We have a chapter by Dennis Fortune from Foster Wheeler on clean-room finishes and materials of construction. This is an intensely practical work, providing information that is difficult to find in any published regulatory or other standard source. It emphasises the importance of an integrated design approach to the selection of finishes and materials of construction, while at the same time frequently referring back to cost control and practical operability.

Two chapters approach contamination control in pharmaceutical clean rooms from more of a laboratory angle. They remind us that contamination control has a principally microbiological focus, and that all forms of microbiological monitoring ultimately rely on competent and knowledgeable laboratory practices and personnel.

Norman Hodges from the University of Brighton, U.K., writes about bioburden determination. John Thompson (Lord Kelvin of Largs) is reputed to have said something along the lines of "When you can measure what you are speaking about and express it in numbers, you know something about it, but when you cannot measure it in numbers your knowledge is of a meagre and unsatisfactory kind."

While not wishing to appear to be in dispute with a long-dead great of world science, this view is not necessarily true of microbiology. Numbers (concentrations) of microorganisms in pharmaceutical products and starting materials and in intermediates are important but so too are the types of microorganisms present. One colony of *Pseudomonas* (*Burkholderia*) *cepacia* per gram in a drug product might be of more consequence than 25 colonies per gram of *Bacillus cereus*. Norman's work emphasises that bioburden has a meaning which, although sometimes forgotten, embraces both numbers and types.

The chapter by Stewart Green and Christopher Randell of Wyeth covers rapid microbiological methods. The traditional means of monitoring the microbiological end-product of the physical, engineering and personnel systems that actually control contamination gives results that come just too late after the time the sample was collected, often four or five days later. This is not just an irritation; current regulatory thinking is placing more and more emphasis on environmental microbiological monitoring, particularly in critical areas of aseptic clean rooms, to the extent that it is known that in at least one major pharmaceutical company more batches of sterile drug products have been rejected for environmental microbiological noncompliances in recent years than have been rejected for failures in the Sterility Test. This is because two or three more product batches may be made on the same line with the same microbiological problem in the period between the problem first arising and its being detected some days later. Significant progress has been made in recent years in developing quicker methods of getting microbiological results. Their application in environmental monitoring and contamination control is still in its infancy. Stewart and Christopher's work brings the reader up to date on the various types of techniques that are becoming available, the scientific principles that underpin them, and gives pointers to the practicalities and limitations of each.

Finally, I hope that somewhere in this book you find something new, that there is something that will be of benefit to you and, for those of you working in the pharmaceutical sector, that there is something that will be of benefit to the company or organisation that employs you. Enjoy.

Nigel Halls

Publisher's Note

Since late 2003, the Medicines Control Agency (MCA) changed its name, after merging with the Medical Devices Agency, to Medicines and Healthcare Products Regulatory Agency (MHRA).

Chapter 1

Effects and Causes of Contamination in Sterile Manufacturing

Nigel Halls

CONTENTS

0-849-32300-2/04/$0.00+$1.50

INTRODUCTION

There is global recognition that pharmaceutical products must always be

- Effective for the therapeutic purposes for which they are prescribed
- Free from side effects that could make them unsafe to use
- Free of chemical, physical or microbiological contaminants that may adversely affect their efficaciousness and safety

The purpose of this chapter is to underpin the principles of contamination control in the manufacturing of *sterile* compounds, by addressing the ways in which pharmaceutical products may be contaminated by microorganisms, materials of microbiological origin and by visible nonviable physical particles. Chemical contamination and cross-contamination are not addressed in this chapter.

Microbiological contamination is not necessarily a problem *per se*. We inhale microbiologically contaminated air when we breathe, we eat contaminated food when we eat, we touch microbiologically contaminated surfaces everywhere. Microbiological contamination is only a problem when it results in unwanted effects caused by contaminated substances, and/or to the user of contaminated substances.

For both sterile and nonsterile pharmaceutical products, the severity of the effects of microbiological contamination is very much a function of the nature of the contaminated product, its intended use, and the nature and numbers of contaminants. At one end of the spectrum microbial contamination of injectable products may lead to death of the patient; at the other end patients may refuse to begin or complete a course of oral medication because of aromas, off-flavors or discolorations of microbiological origin.

1 EFFECTS

1.1 Introduction

Limited microbiological contamination is tolerated in nonsterile pharmaceutical products such as inhalations, tablets, oral liquids, creams and ointments, etc. The pharmacopoeias and the regulatory bodies responsible for licensing the manufacture of pharmaceuticals may require the numbers of microbiological contaminants per unit volume or weight of these products to be limited, and that specified microorganisms are restricted throughout product shelf life. Compliance with these limits is, in most cases, sufficient to protect the patient from unwanted adverse effects.

The microorganisms for which there are specific restrictions in nonsterile products are only indicators of types that could cause infections when the drug product is used as directed. Naturally there should be no pathogens present, but pharmacopoeial monographs primarily exist as standards against which products are tested. It is recognized that it would be impractical — with existing methodologies — to test pharmaceuticals exhaustively for all potentially pathogenic contaminants.

In the *United States Pharmacopeia (USP) XXVI 2003*, there are only 95 monographs that include microbial limits. Fifty-one of these require absence of *Staphylococcus aureus* and *Pseudomonas aeruginosa*, 20 g or 10 ml, 20 require absence of *Escherichia coli* **or** *Salmonella* spp., or *Escherichia coli* **and** *Salmonella* spp., from 10 g or 10 ml. The restrictions on *S. aureus* and *P. aeruginosa* apply to topical products, because these microorganisms are typical of types that could cause infection when products are used on open wounds or abraded skin.

The restrictions on *Escherichia coli* and *Salmonella* spp. are applicable to oral products because these microorganisms are typical of types that could cause gastrointestinal infections.

The pharmacopoeial restricted species have been chosen as indicators, at least in part, because of the availability of robust techniques for their isolation and recognition. The possibility of other objectionable microbiological contaminants in nonsterile products cannot be disregarded.

When contamination is discovered, its significance must be evaluated conservatively, considering the formulation of the product, its method of delivery, the contaminant, and the type of patient undergoing treatment. For instance, in 1994 a U.S. company responsibly and voluntarily withdrew 3.6 million units of albuterol sulfate inhalation solution from the market on confirmation of contamination with *Pseudomonas fluorescens*. *Bergey's Manual of Determinative Biology* recognizes *Pseudomonas fluorescens* as being more likely to be associated with soil and water than with specific pathogenicity to humans. A team of independent microbiologists set up at the time of the recall concluded that *Pseudomonas fluorescens* has "very rarely been found to be the causative agent of illness."

The reason for the recall was concern that this microorganism could cause lung infections, which could be particularly serious in people with cystic fibrosis, chronic obstructive lung disease or with compromised immune systems.

Nonsterile pharmaceutical products are generally formulated to prevent any microorganisms from increasing in number during their shelf lives. This may be intrinsic to the dosage form. An example in solid dosage forms, such as tablets or powder inhalations, is the lack of sufficient water to allow microorganisms to multiply over time. Conversely, nonsterile aqueous dosage forms, in which there is sufficient water to potentially allow microorganisms to multiply, are usually formulated to incorporate antimicrobial preservatives.

In addition to these formulation-related factors, there are regulatory requirements governing the standards of hygiene applicable to the manufacture of nonsterile pharmaceuticals. Such regulations may restrict the numbers and types of microbial contaminants that could be initially present on the product (i.e., at release as distinct from at the end of shelf life). The required standards of hygiene, although exacting and often involving filtration of environmental air, do not normally require manufacture of these products in clean rooms, in the sense that the term "clean room" is understood in the sterile products manufacturing industry.

1.2 Contamination of Sterile Products

Sterility is defined as "freedom from all viable life forms."

Two broad groupings of pharmaceutical products are required to be sterile — parenteral and ophthalmic products. Such products must be free from all viable life forms, due to the potential consequential severity of the consequences of viable microorganisms present when the products are used in the manner intended or prescribed.

Confirmed incidents of nonsterility in supposedly sterile parenteral and ophthalmic products have been comparatively rare. However in the 1970–1971 Rocky Mount incident in the U.S., 40 deaths were attributed to nonsterile infusion fluids contaminated by *Enterobacter cloacae*, *Enterobaccter agglomerans* and other *Enterobacter* spp. (Felts et al., 1972; Maki et al., 1976). In the 1971–1972 Devonport incident in the U.K., five deaths of postoperative patients were attributed to nonsterile dextrose infusions contaminated by *Klebsiella aerogenes* (Clothier Report, 1972).

In the 1972–1973 Chattanooga incident in the U.S., three deaths were attributed to *Enterobacter cloacae*, *Enterobacter agglomerans* and *Citrobacter freundii* (CDC, 1973). In each incident there were many more nonfatal bacterial septicaemias. More recently there were 46 cases of bacterial septicaemia in Spain attributed to a nonsterile *Burkholderia (Pseudomonas) pickettii* contaminated aseptically filled ranitidine injection (Fernandez et al., 1996).

In 1964, eight patients in Sweden developed postoperative eye infections caused by *Pseudomonas aeruginosa*-contaminated eye ointment — one of the victims was left blind (Kallings et al., 1966).

More recently, in November 2002, the FDA issued a nationwide alert on all injectable drugs prepared by Urgent Care Pharmacy in South Carolina, based on lack of assurance that their products were sterile. A 77-year-old woman died and two other patients contracted an extremely rare fungal meningitis after receiving spinal injections of methylprednisolone prepared by Urgent Care. Spinal fluid from the patients tested positive for a rare fungus consistent with that found in the Urgent Care product.

1.3 Parenteral Products

Parenteral products are intended for administration by injection, by infusion, or by implantation into the human body. Products normally totally free from microbiological contamination or colonization are delivered to internal tissues, while the parenteral route of administration deliberately bypasses the body's external physical barriers to infection.

No distinction can be made between microorganisms known to specifically cause infectious disease in humans, from those customarily thought to be harmless or benign. Once the body's external defensive barriers have been broken down, it is reasonably conservative to assume that any microorganism may potentially find nourishment in internal tissues, thereafter proliferating and causing infection.

Virtually *any* microorganisms can cause infections in immunosuppressed or immunodeficient patients. None of the four bacterial species (classed according to *Bergey*) associated with the fatalities of the 1970s (*Citrobacter freundii*, *Enterobacter agglomerans*, *Enterobacter cloacae*, *Klebsiella aerogenes*) are thought to be more than "opportunistically" pathogenic, and all may be found living in commensal association with healthy humans. Commonly found skin bacteria such as *Staphylococcus epidermidis* are not unusually found in postoperative infections.

1.4 Viability

Viability is defined as the capability of microorganisms to grow, divide and increase sufficiently to form visible colonies on solid nutrient media, or turbidity or other visible change in fluid nutrient media. Within the need for sterility in parenteral products, the presence of one viable microorganism in a sealed product unit is considered sufficient to potentially cause infection.

However, this is only partially true. The numbers of viable microorganisms in lethally contaminated infusion fluids during the 1970s were, in all cases, in concentrations exceeding 10^6/ml; several hundred millilitres were possibly infused into each patient.

The presence of foreign matter in the body or at the site of injection is known to influence the threshold number of microorganisms required to cause a clinically recognizable infection. Elek and Conen (1957) established that when *Staphylococcus pyogenes* alone was injected into human volunteers, 10^6 microorganisms were required to produce a pus-forming infection, but only 10^2 were required when foreign matter (braided silk suture) was included with the inoculum.

The apparent requirement for a threshold number of microorganisms, which must be exceeded to overwhelm patient defence mechanisms and cause infections, has been confirmed experimentally for *Staphylococcus aureus* and Gram-negative bacteria.

Other microorganisms may survive in small numbers in the human body; for instance, *Streptococcus viridans* may generate a protective slime and adhere to diseased natural tissues. Subsequently, perhaps during periods of immunodepression, it may proliferate and establish infections (Dougherty, 1988).

The pharmaceutical industry regards even one viable contaminant in a sealed product unit as a compromise of sterility, regardless of the size of the product unit, whether it be a 0.5-ml subcutaneous injection or a 1litre intravenous infusion.

1.5 Pyrogens

Sterility is not the only attribute related to contamination significant in parenteral products. Parenteral products have to meet limits on pyrogens and on nonviable particles.

Pyrogens are substances which, when injected in sufficient amounts into the mammalian body, will cause a rise in body temperature. Most deaths arising from the contaminated infusion fluids of the 1970s were caused by pyrogenic shock, rather than as a result of bacterial septicaemia.

The most significant source of pyrogens is microbiological, specifically lipopolysaccharide fractions of the cell envelope of Gram-negative bacteria. These substances are referred to as bacterial endotoxins. Practically, the term bacterial endotoxin can be regarded as synonymous with the term pyrogen.

1.6 Endotoxins

Endotoxins have molecular dimensions small enough to pass freely through bacteria-retentive filters. They are also heat stable at steam sterilization temperatures. Typically, endotoxicity is not lost with loss of viability. Treatments that are intended to achieve sterility may not guarantee freedom from pyrogenicity if the product was heavily contaminated prior to sterilization.

Pharmacopoeial limits on endotoxins in sterile parenteral products are calculated from a formula that takes account of the concentration of endotoxin in the product, the dosage regimen and the weight of the patient. The formula is expressed as K/M, where:

K = the approximate threshold pyrogenic dose for humans. With some exceptions this has been given a fixed value of five Endotoxin Units (EU) per kilogram of body weight of the patient (70 kg is used in calculations as the weight of an average human patient)

M = the maximum dose of product per kilogram of body weight of the patient that would be administered in a single one-hour period.

The EU relates back to the first batch of the *USP* Reference Standard, which contained one EU per 2×10^{-8}g of the standard endotoxin. The threshold pyrogenic dose would be in the order of 10^{-9} g per kilogram of body weight of the patient. Since endotoxins occur in Gram-negative bacteria to the extent of about 10^{-15} g per bacterium, this is equivalent to requiring injection or infusion of about 10^6 bacteria per kilogram of body weight of patient to induce a pyrogenic reaction, or 7×10^7 bacteria (living or dead) for the average 70-kg patient.

1.7 Particulate Matter

Particulate matter has, for parenteral products, been defined as "mobile undissolved substances which are unintentionally present." It is divided into subvisible and visible particles with the limit at 50 μm. Intravenously administered pharmaceutical products enter the circulatory system and pass through the lungs, where the largest particles are filtered out before the product is pumped on through the arterial circulation.

The potential for patient risk from nonviable particles was first reported in the 1950s, gathering impetus when it was demonstrated in the 1960s, that foreign body granulomas could be produced in the lungs of rabbits following administration of commercially available parenteral infusion solutions.

Thrombosis and phlebitis are clinical complications for which there is sound documentary evidence of both conditions being caused by nonviable particulate contamination (Akers, 1987) of parenterally administered pharmaceutical products.

The sources of nonviable particulate contamination of parenteral products is divided into *intrinsic* and *extrinsic* origins (Backhouse et al., 1987). Intrinsic contamination comes from the areas of manufacture, packaging, transit, and storage. Extrinsic particulate contamination is introduced at the time of drug reconstitution and usage.

Most intrinsic nonviable particulate contamination of parenteral products is thought to originate in

- Product-contact packaging materials
- Leaching and dissolution of the surfaces of glass containers (flaking)
- Rubber closures (Desai, 1987)

The manufacturing environment may, unless controlled carefully, be another important source:

- The use of aluminium for transport containers or for wall finishes
- Machinery used in parenteral manufacture, a well-known intrinsic source of nonviable particulate contamination

1.8 Ophthalmic Products

Ophthalmic products must always be sterile, because the cornea and other transparent parts of the eye are extremely susceptible to irreversible loss of transparency as a result of microbiological infection because they have a particularly poor immune response due to a low blood supply.

Pyrogenicity is of no relevance to ophthalmic products. It is large particles that carry risk of physical damage to the eye. Most eye drops and eye ointments are supplied in multidose, nonresealable containers. Sterility of the contents of these containers is compromised as soon as they are opened. This may be an argument that ophthalmic products are not truly sterile in the same sense that parenteral products are. Nonetheless, they are required to be sterile, and must be manufactured to the same stringent standards of sterility as parenteral products.

Sterility of an ophthalmic product in use is achieved by inclusion of antimicrobial preservatives in their formulations. The inclusion of preservatives is not intended to chemically sterilize the product in manufacture, only to inactivate contaminants that may arise in use.

Typically, these products are allocated two shelf lives. The first, often measured in years, applies to the product while its container is still sealed; the second, usually measurable in weeks, applies after the container is opened. This recognizes the limitations of preservatives used in ophthalmic products in relation to the ability of some microorganisms, given sufficient time, to develop resistance to antimicrobial preservatives.

1.9 The U.S. Requirement for Sterility in Aqueous Inhalations

In May 2000, the FDA amended its regulations requiring that all aqueous-based products for oral inhalation be manufactured sterile. The FDA rationalized that such was the danger of nonsterility to patients with cystic fibrosis, coupled with the fact that most aqueous inhalations were already being manufactured as sterile, that a rule might as well be put in place.

The contradiction inherent in this rule is that the systems for delivery of inhalation products to the patients (nebulizers) are not required to be sterile.

2 CAUSES OF CONTAMINATION

2.1 Introduction

Microorganisms are ubiquitous. In nature, potential sources of microorganisms are literally limitless. However, the huge array of potential sources of contamination is severely restricted when indoor pharmaceutical manufacturing environments are considered.

There are four major sources of microbiological contamination: air, water, materials and people.

A good working knowledge of how contamination may arise from these four sources, and how they can be controlled, is essential to the successful design and operation of microbiologically controlled environments and clean rooms.

2.2 Contamination from Air

Air is probably more frequently regarded as a vector of microbiological contamination than as a primary source. It is not a nutritive environment; microorganisms do not grow and multiply in air. Many microorganisms die in air. Anaerobes die as a result of oxygen toxicity but, more generally, aerobes die as a result of desiccation. Photosensitivity may also play a part in inactivating certain bacteria in air.

All natural air is microbiologically contaminated. Most microbiological contamination is associated with nonviable particles in the air such as dust, skin flakes, etc. Nonviable particulate matter is both a source of microorganisms and also a means of protecting microorganisms from death by desiccation.

The most likely types of microorganisms traceable to air are those with mechanisms to resist desiccation such as *Bacillus* spp., *Micrococcus* spp. and fungal spores, which have evolved to be dispersed in the air.

The most likely sources of *Bacillus* spp. are from excavation or building work, where soil or dust is disturbed. *Micrococcus* spp. may also survive in soil or dust, though they are more likely to be of human or animal origin.

The primary means of controlling airborne contamination in pharmaceutical manufacture is by the use of clean rooms or, in more critical cases, isolation technology. The manner in which clean rooms must be designed and operated effectively also places restrictions on contamination from other sources such as people, materials and water. In clean rooms and in isolators, contamination from air is controlled by a number of different mechanisms — based on filtration, dilution, pressure differentials and air flows. Further detailed information on these may be obtained from standards such as IS 14644 (*Cleanrooms and Associated Controlled Environments*) and in general texts such as those of Whyte (1991) and Wagner and Akers (1995).

2.3 Clean Rooms Defined

To merit the term “clean room,” an internal area must be supplied with filtered air.

The types of filter generally used for this purpose are not classifiable as sterilizing filters, and would not pass the stringent tests applicable to bacteria-retentive sterilizing filters. They are capable of removing the great majority of microorganisms. When coupled with other filters, and with recirculation of previously filtered air to dilute the challenge of air from uncontrolled sources, they can be extremely effective in the maintenance of asepsis.

Clean rooms must be maintained at higher differential air pressures than adjacent uncontrolled areas. This prevents the quality of air achieved by filtration and dilution being compromised by any unsealed or openable junctures with lower standards in adjacent areas. The result of the pressure differentials is a constant outward air flow at unsealed points, and outflow for limited periods when doors are opened. Microorganisms are unable to move against pressure differentials, and in the opposite direction to air flows. The limits for pressure differentials and air flow values normally used in clean-room design and control are based largely on history rather than scientific data.

The inability of microorganisms to "swim against the tide" is also relevant to the use of laminar flow air to protect specific areas in sterile pharmaceutical manufacture. Filtered air is accelerated across the area, and is protected so as to prevent back flow, mixing and turbulence. Air moving in this manner is described as *laminar flow (or unidirectional) air*.

Laminar flow air serves two purposes. First, it provides a protective directional air movement, preventing microorganisms entering the protected area. Second, it is able to "sweep away" microorganisms already in the area. This second attribute is somewhat arguable, and whereas there is no doubt that laminar flow air (such as in air showers) can clean microorganisms from surfaces, etc., it is rarely used as the sole means, and is never used to clean microorganisms from surfaces and materials unless they have been precleaned by other means.

2.4 Contamination from Water

Water is a very serious source of microbiological contamination. Microorganisms, particularly Gram-negative bacteria, can grow and multiply in water, even when nutrients are only present in very low concentrations. Some such microorganisms evolve to form films or slimes, which adsorb nutrients from flowing water. Periodically these films are naturally sloughed off into the water stream.

Most microorganisms are unable to move or expand more than a few millimetres on dry, solid surfaces. Conversely, waterborne types are guaranteed to be found in practically every wet location, and in locations that have recently been wetted. Water is therefore a vector, as well as a source, of microbiological contamination.

The most likely types of microorganisms traceable to water are *Enterobacteriaceae*, including *Escherichia coli* and *Salmonella* spp. as the most readily recognizable types in domestically contaminated waters, and *Pseudomonas* spp., which are always found in natural, potable and pharmaceutical waters.

There are regulatory restrictions on the presence of water in clean rooms used for aseptically filling sterile pharmaceutical products. The restrictions are — on the face of it — quite simple: there should be no water outlets in these areas.

However, water is the most commonly used cleaning fluid and diluent for pharmaceutical products, cleaning agents, disinfectants. It is virtually impossible to have a completely water-free pharmaceutical clean room, and it is certainly impractical. Water-based fluids must be sterilized before entry to aseptic filling rooms, usually by filtration.

Water is permitted, and is indeed necessary, in the lower grades of clean rooms (change rooms, preparation areas, compounding areas, etc.), which surround and exist to service sterile pharmaceutical manufacture. Wherever possible it should be of pharmaceutical grade (of purified water or water for injection quality), microbiologically controlled and monitored. Personnel movement from areas in which there is water into high-grade, aseptic clean rooms may be an additional vector for waterborne contaminants.

2.5 Contamination from Materials

All materials brought into a microbiologically controlled environment are potential sources of contamination. Any microorganisms can be associated with undefined materials. It is impossible to make a general assessment of risk from these sources, except to distinguish materials of plant, animal or silicaceous origin as being more likely sources of contamination than materials produced by chemical synthesis.

It is good practice for all materials and their manufacture to have been evaluated by audit (for their potential to contaminate pharmaceutical manufacturing clean rooms), but it may not be practical or economically possible to avoid contamination from such sources. Microbiological monitoring programmes should be operated to ensure that the pharmaceutical manufacturing facility is not exposed to the worst excesses of these potential sources of contamination. Suspect materials should be monitored on the basis of every incoming batch, but this need not necessarily apply to all materials, particularly synthetic chemicals.

Contamination of materials in transit and warehousing should be considered. Water and other damage to external packaging is particularly relevant, and should be referred to the department with the expertise to make a professional analysis of the risk to the product. This is usually Microbiological Quality Assurance (MQA).

2.6 Contamination from People

People are a significant source and the most unpredictable vector of microbiological contamination. Microorganisms are always present on hair and skin, which are shed into the surrounding environment. With more movement, more microorganisms are shed.

Concentrations of microorganisms are found in the nose, throat, mouth, anal and genital regions, and may be dispersed by breathing, coughing, sneezing, talking, flatulence and hand contact.

The most likely types of microorganism traceable to natural shedding are

Staphylococcus spp. and *Micrococcus* spp. *Propionibacterium* spp. and other coryneforms are unquestionably of human origin but, although periodically identified in failed sterility tests, are hardly ever isolated in environmental monitoring programmes. This may be due to their sensitivity to oxygen and to light limits, the length of time they can survive in air, but most likely by direct transfer from personnel without air being involved as a vector.

Staphylococcus spp. and *Streptococcus* spp. may be traceable to the nose and throat, *Enterobacteriaceaea* to the anal and genital areas. Yeast may be traceable to nose, throat or genitals.

Contamination from people is generally controlled by two means.

"Packaging the Personnel"

Personnel who work in clean rooms must be provided with garments suitable for the type of work and clean room. In high-grade, aseptic clean rooms this usually means that all garments are sterile, made from bacteria-retentive fabrics, nonlinting, and leaving as little as possible uncovered skin. In support areas it is unnecessary to have such stringent garment control. No matter how severe the restrictions placed on garments, they are a compromise — personnel have to breathe, perspire, move, see, hear, and so on.

The logic of particular restrictions is always challengeable: "Why do I have to wear a head cover when I shave my head each day? Look, that guy over there has bushy eyebrows and you don't ask him to cover them!"

Informed common sense should prevail.

Training (Education)

Most people are well intentioned. When they know that there is a correct way of doing things, they usually do it that way. When they understand the reason behind a particular way of doing things, they are even more likely to do it in the proper way. Managers in the pharmaceutical industry are responsible for training (educating) personnel in asepsis, in proper ways of changing into their clean-room garments without contaminating them, the change rooms, the clean rooms, and in proper behaviour in the clean rooms.

Some pharmaceutical manufacturing companies claim to restrict personnel from clean rooms if they have been shown to carry pathogenic *staphylococci* or *streptococci* in their throats or noses. It is very important that personnel with symptomatic medical conditions leading to excessive shedding or dissemination of microorganisms, e.g., coughs, colds, flaking eczema, etc., be restricted from clean rooms.

The attempt to restrict nonsymptomatic carriers is a different issue of complexity. Why should nonsymptomatic carriers of *Staphylococcus aureus* in their noses be more of a risk to sterile products than nonsymptomatic carriers of *Staphylococcus epidermidis* on their skins (as most people are)?

2.7 Modeling Contamination

It is important to understand how microbiological contamination occurs, and how materials become nonsterile in the creation of clean rooms. Various agencies, notably the National Aeronautics and Space Administration (NASA) (Hall, 1965; NASA, 1968; Hall and Lyle, 1971), have interested themselves in microbiological contamination and in developing models for how it arises. This has increased in importance when the potential contamination of other planets with life-forms originating from earth is considered.

2.8 The Plateau Effect

The most important observation underpinning our understanding of contamination is called the "plateau" effect (Roark, 1972; Sykes, 1970). If an inert surface is left in a microbiologically contaminated environment, one might reasonably expect a gradual and continuous increase of microorganisms recoverable from the surfaces. This is not the case. The microbial count per unit area increases and then equilibrates (the plateau) for an indefinite period thereafter (Figure 1.1).

Figure 1.1. The plateau effect.

The plateau effect has led to the development of theories of contamination. Figure 1.2 shows a pictorial model of how a sterile item may become microbiologically contaminated when placed in a nonsterile environment. This model proposes two mechanisms of contamination: deposition and contact.

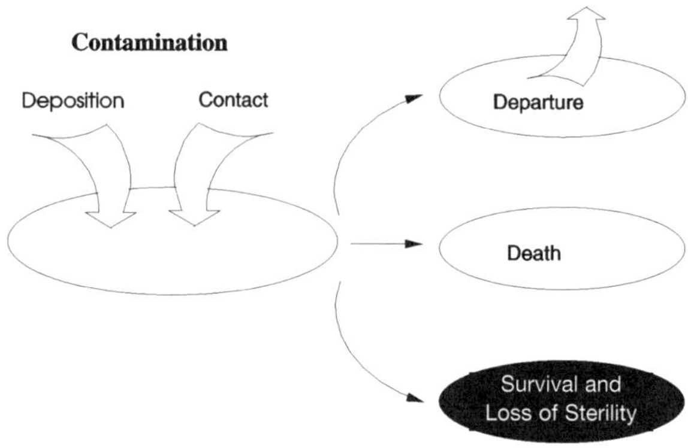

Figure 1.2. Contamination and loss of sterility.

Deposition recognizes that microorganisms present in environmental air are likely to settle out on surfaces of items, as a result of any one or more of several mechanisms. Contact describes contamination by means of transfer of microorganisms from one surface to another by physical proximity. The contribution from these two mechanisms to the contamination of items will differ according to individual circumstances.

The extent of contamination from deposition will differ according to the concentration of microorganisms in air; this will differ from one type of environment to another, and within any one environment it will differ from one time to another. The outstanding feature of the design of pharmaceutical clean rooms used for the manufacture of sterile products is the extent of control of the microbiological quality of environmental air, particularly around areas where the product is exposed.

Air is filtered, often recirculated and refiltered. It is maintained in constant turbulence or is used in laminar flow devices to "sweep" contaminants away from exposed items. In a well-designed process operating in a well-designed, well-maintained clean room, contamination by deposition of microorganisms from environmental air is intended to be controlled to a "steady state," where it is not likely to be a significant persistent mechanism of product contamination.

2.9 Contamination and Loss of Sterility

The vulnerability of product contamination from deposition increases when the steady state is disturbed. Personnel are the most significant cause of disturbance.

The reality of clean rooms is that no matter how skilled, well-trained or well-garbed, the concentration of microorganisms and nonviable particles in air around personnel is inevitably higher than in air in unmanned areas.

If it is accepted that personnel are necessarily present in or around areas where product is exposed, for instance to start up the process, make adjustments, take samples, monitor etc., it should be accepted that the amount of contamination from deposition will then increase.

Bernuzzi et al. (1997) summarized these views by stating that contamination in aseptic filling of pharmaceutical products is mainly the result of two different stochastic processes.

The first contribution to contamination is from airborne particles, while the second is from personnel line intervention. The first spans the whole filling operation, the second occurs randomly when human intervention takes place.

Sometimes asepsis is referred to as a "no-touch" technique, thereby reducing contamination by contact to the minimum. Primary sources are personnel and water, but equipment, machine surfaces and even integral components of the pharmaceutical presentation may be vectors for contact contamination.

Contamination could occur in a pharmaceutical product in a vial from contact with a rubber closure, which has in turn been contaminated by contact with the production operator while transferring the sterilized closures from the autoclave to the hopper on the filling machine. Contamination by contact is intermittent, erratic and largely unpredictable.

The second important consideration illustrated by this model (Figure 1.2) is that contamination is not synonymous with loss of sterility. Sterility is defined as the absence of all viable life forms from an item. Clearly, the plateau effect illustrates that an item may become contaminated, but the fate of its contaminants may thereafter follow three courses, only one of which necessarily leads to nonsterility.

The microorganisms that have been transferred to the item by physical forces may as well be removed by physical forces; they may fall off, fall out or be blown off the item. The microorganisms may die on the item; death rates of microorganisms are particular to species and to the nature of the material they find themselves to be in or on, and to the surrounding environmental conditions. Desiccation-resistant types have a greater potential for survival on inert surfaces.

It is important to understand this distinction between contamination and nonsterility.

Experimental work has been done to develop and support views, theories and mathematical models of aseptic manufacture developed from techniques involving the recovery of microorganisms in liquid nutrient media (media fills), or on solidified nutrient media (active and passive microbiological air monitoring).

The conditions in microbiological media are, with respect to the survival potential of microorganisms, quite different from the conditions existing in "inert" materials used in aseptic manufacture. These include glass vials, rubber stoppers and stainless steel hoppers, as well as aseptically manufactured pharmaceutical products (e.g., nonneutral pH, antimicrobial preservative content, hypotonicity, hypertonicity, etc.).

2.10 Mathematical Models

Roark (1972) developed the following model to describe the microbiological contamination of spacecraft. Its principles are applicable to any form of contamination, including the manufacture of sterile pharmaceutical products.

Number of contaminants after time $t = A \, . \, \lambda(t) \, . \, \mu(t) \, . \, q_i$,

where

A = the surface area exposed to contamination. The larger the area the greater the probability of microbiological contamination

$\lambda(t)$= the deposition rate of microbial contaminants on the item of surface A. The symbol (t) describes the elapsed time in which the item is exposed to the contamination potential.

$\mu(t)$ = the removal rate of microbial contaminants through physical means or death. The symbol (t) describes elapsed time. The proportion of survivors within a microbial population diminishes as a function of time.

q_i = the number and manner in which microbial contaminants may be present in the contaminating environment whether as individual bacteria ($i = 1$) or as groups or clumps borne on nonviable physical particles. The symbol i represents the number of microbial contaminants (0, 1, 2, n) that may be present in any one clump. There is extensive microbiological evidence to indicate that in nature many viable airborne microorganisms are attached to larger, nonviable particles such as skin flakes, dust, lint. The size and composition of these larger particles influences the ease or difficulty with which the particles will settle out of air and also the ease or difficulty of their physical removal from contaminated items (e.g., due to their very small size, discrete microorganisms are very difficult to remove, but conversely, they are not protected from desiccation by any extrinsic factors).

The mathematical complexity of this model is unimportant. It is consistent with observations, and identifies the factors that must be resolved in order to describe the contamination processes. None of the functions are resolved in this model to the point where it could be applied to pharmaceutical manufacturing.

Bradley et al. (1991) studied contamination in a containment room, where they established uniform, stable concentrations of 10^4, 10^6 and 10^7 discrete airborne spores per cubic millimetre (mm^3) of *Bacillus subtilis* var *niger*. The test system was a blow-fill-seal machine filling Tryptone Soy Broth (TSB) at a fixed rate into plastic ampoules.

They demonstrated a regular relationship between the logarithm of the fraction of product units contaminated and of the spore challenge concentration in air. The team claimed that, by extrapolation of this relationship, they could substantiate a sterility assurance level (SAL) for practical operating conditions in actual airborne contamination. They stressed that the observed relationship was specific to *the experimental conditions*.

The regularity of the form of the relationship described by Bradley et al. (1991) is observationally important as an unconfused, assumption-free description of how airborne contamination relates to product contamination. It is, however, only a measure of the contamination frequency.

In terms of Roark's (1972) model it only describes the deposition rate $\lambda(t)$ and does not extrapolate to SAL. It does not take contamination by contact into account (perhaps understandably, in that the blow-fill-seal process affords little opportunity for contamination by contact than other aseptic filling processes).

By using a TSB recovery system, the findings disregard the product-specific die-off of microorganisms in filled units.

Within its description of deposition rate, the result only accounts for contamination by discrete microorganisms. Natural patterns of contamination from airborne sources would be much more complex, and should consider the size and nature of nonviable clumps.

The general form of the log–log relationship in these studies could form the basis of a more complex model for what Bernuzzi et al. (1997) described as the "background" contamination from airborne sources, which operates throughout an aseptic filling process.

2.11 Whyte's Analyses

Throughout the 1980s and 1990s William Whyte and his co-workers attempted a more ambitious model of contamination than the experimentally based views on deposition published by Bradley et al. (1991).

In 1986 Whyte listed five mechanisms by which airborne particles can be deposited on surfaces. He analyzed in some detail the significance of each of these mechanisms to contamination in practice in pharmaceutical clean rooms.

Whyte's analysis is based on common sense, observation and experience, coupled with some practical experimentation.

Whyte's five mechanisms are:

1. *Brownian Motion.* As this factor is only applicable to particles of 0.5 μm or smaller, Whyte concluded that it would be of no practical significance in pharmaceutical clean rooms. This would have been about the size of the discrete microorganisms used in the experimental system of Bradley et al. (1991). Whyte argues that airborne microorganisms are actually carried on much larger particles and references 14 μm for the typical size of airborne particles from hospitals (Noble et al., 1963), 20 μm for the median size of skin

flakes (MacIntosh et al., 1978), and 7, 10, 11 and 17 μm (varying according to garments worn) for bacteria-carrying particles shed by workers in pharmaceutical clean rooms (Whyte, 1984, 1986).

2. *Inertial Impaction.* In 1981 Whyte presented mathematical and experimental evidence concerning the effects of particulate contamination of bottles through their open necks as a consequence of gravity and inertial impaction. These are quite different according to whether the air stream around the open neck of the bottle is at right angles, or parallel to the neck. Impaction made a greater contribution to contamination when the air stream is parallel to the neck. Mathematical models presented in this publication predicted that contamination by inertial impaction should be of similar importance that by gravitational settling for microorganisms and for nonviable particles in the size range of 5 to 20 μm.

   In 1986 Whyte contended that he had previously overstated the importance of impaction in order to present a worst-case scenario, and that in actuality it would be less significant than gravitational settling.

   Conversely, it is possible that inertial impaction could account for the significant effects of laminar air flow protection (on or off) on the position (but not the general form) of the relationship between the concentration of airborne microorganisms and the frequency of contaminated blow-fill-seal ampoules reported by Bradley et al. (1991).

   Some factoring for inertial impaction merits inclusion in an expansion of Roark's (1972) term $\lambda(t)$, the deposition rate. Whyte's (1981) expression for the number of particles impacted in time T is probably as good a basis as any.

$$\text{Number of particles impacted} = C\,.\,A\,.\,V\,.\,P\,t,$$

   where

   C = the concentration of airborne microorganisms.

   A = the surface area exposed to contamination. Whyte (1981) presented this as the diameter of the open neck of a bottle, but it could equally apply to the surface area of a rubber vial stopper, etc.

   V = the velocity of the air carrying the microorganisms or particles. Impaction has its greatest influence on rapidly moving particles.

   P = an inertial parameter defined by the shape of the item upon which impaction may take place,e.g., cylindrical, spherical, etc.

   t = the elapsed time in which the item is exposed to the potential of contamination.

3. *Direct Interception.* Van der Waal's force attracts particles onto surfaces when the two are very close together. Whyte (1986) discounted any significant contribution from these forces of direct interception to contamination of air flow-protected surfaces in clean rooms. It is difficult to see how they could have a major effect in environments of continuously turbulent or fast-streaming air, with only low concentrations of contaminants present.

4. *Electrostatic Attraction.* These forces can operate at much greater distances than can the forces of direct interception. They depend on the electrostatic charges present on materials. Glass containers are likely to have very little charge and have less electrostatic attractiveness than more highly-charged plastic containers. The fabric of clean room operators' garments should be chosen carefully to ensure that electrostatic charges do not build up until the operator becomes a "magnet" for airborne particles, which he may then transfer by direct contact to the product or to product contact components (e.g., vial stoppers). The choice of materials used in clean rooms and for clean room clothing and furnishings virtually eliminates electrostatic forces causing a real problem.
5. *Gravitational Settling.* This reflects Whyte's 1986 thesis, and relies on gravitational settling being the principle means of deposition of microorganisms in clean rooms. This thesis presented a model to approximated deposition by means of Stokes Law in which the settling velocity of particles in fluids are described by the following equation:

$$V_S = \rho \cdot g \cdot d^2 / 18\gamma,$$

where

V_S = settling velocity of particles in a fluid

ρ = the density of the particle. The density of skin flakes and similar particles which carry airborne bacteria (this can be taken to be equal to one).

g = the acceleration due to gravity.

d = the diameter of the particle(s) in the air. Whyte (1986) used a diameter of 12 µm in subsequent predictive calculations. This is an approximation: there is sufficient experimental evidence to indicate that there may be quite a range of sizes of particles carrying microorganisms in air.

γ = the viscosity of the fluid within which the particles are settling. Air can be assumed to have a viscosity of 1.7×10^{-4} poise.

From this equation and the assumption that microorganism-carrying particles are of 12-µm equivalent diameter, Whyte (1986) concluded that their settling rate in air is 0.462 cm/sec. Sykes (1970) alleged that the settling rate in air calculated by Stokes Law for particles of 5-µm equivalent diameter is about 0.07 cm/sec, some six or seven times slower than Whyte's (1986) figures. The difference is probably due to different assumptions within the application of Stokes Law. Whyte (1986) contended that measurable contamination rates can be predicted by Stokes Law, because gravitational settling is the principle cause of contamination in clean rooms. Sykes (1970) contended that the greatest risk of contamination in clean rooms comes from "moving air carrying microorganisms in the direction of, or onto, the sterile surface," in other words *inertial impaction.*

Undoubtedly gravitational settling must play some part in deposition, and Stokes Law should take its place alongside the equation describing inertial impaction in any expansion of Roark's (1972) term $\gamma(t)$, the deposition rate. There is no experimental evidence to substantiate the balance of the two factors and how they may be affected by physical conditions.

Whyte et al. (1982) conducted a series of practical experiments in two semi-automated aseptic filling rooms to obtain four different contamination rates for hand-stoppered, TSB-filled vials under four different conditions of airborne contamination. The actual contamination rates obtained in these experiments were compared (Whyte, 1986) with theoretical contamination rates derived from the Stokes Law thesis on gravitational settling using measures of airborne contaminated from settle-plate data, and from volumetric air sampling. Corellations were not good, seven of the eight predicted rates were higher than the actual rates of contamination. Whyte (1986) contended that the predictions were good estimates, erring on the conservative side.

It is possible to conclude that Whyte overemphasized the importance of gravitational settling.

Deposition has dominated the interest in contamination modelling. In modern, well-controlled clean rooms it is probably a very minor component in product contamination, especially when compared to contamination by contact.

However, contact is an even more difficult concept. Bernuzzi et al. (1997) used the term "outliers" to describe incidences of contact contamination. Contact contamination is likely to be an intermittent factor, and may not be confined to point of fill, or to the time frame in which a filling operation is conducted. For instance, it is possible for rubber vial closures to be contaminated when they are unloaded from the autoclave one day and filled on another.

Later it is possible for that contamination to be redistributed among the closures when they are transferred to the hopper, eventually to randomly contaminate product units when the closures are pushed home.

The potential importance of contact contamination (hand-carriage contamination and the protective effects of clean-room clothing) was illustrated by Whyte et al. (1982) and by Whyte and Bailey (1985). There was a ten-fold difference in contamination rates of TSB-filled, hand-stoppered vials between operators wearing isopropyl alcohol-disinfected gloves and those with unwashed bare hands.

Roark's (1972) factor $\mu(t)$ describing the removal rate of microbial contaminants through physical means or death has only been addressed in terms of microbial death. The general form of microbial death is known to follow an exponential form (Fredrickson, 1966). Whyte et al. (1989) showed with a wide range of parenteral products that most were unable to support the growth or survival of any microorganisms, except for a few Gram-negative types in mainly unpreserved products. Physical removal is a largely undocumented topic.

In conclusion, contamination modeling as it applies to aseptic pharmaceutical manufacture in clean rooms is still in its infancy. The mechanisms are clearly complex and probably unique to each facility and filling operation, and to their air-flow protection, manning, clean-room garments and disciplines.

Experimental data arc difficult to generate, and the assumptions supporting particular models may not bc transferable from one situation to another. Contamination as measured by growth in nutrient media should not be considered synonymous with the assurance of sterility (SAL) for particular pharmaceutical products; at best it is a worst-case, but grossly inaccurate, estimate of SAL.

REFERENCES

Akers, M.J. Current problems and innovations in intravenous drug delivery. *American Journal of Hospital Pharmacy*, 44: 2528–2532, 1987.

Backhouse, C.M., Ball, P.R., Booth, S., Kelshaw, M.A., Potter, S.R., McCollum, C.N. Particulate contaminants of intravenous medications and infusions. *Journal of Pharmacy and Pharmacology*, 39: 241–245, 1987.

Bergey's Manual of Determinative Bacteriology, 9th ed. Baltimore and London: Williams & Wilkins, 1994.

Bernuzzi, M., Halls, N.A., Raggi, P. Application of statistical models to action limits for media fill trials. *European Journal of Parenteral Sciences*, 2: 3–11, 1997.

Bradley, A., Probert, S.P., Sinclair, C.S., Tallentire, A. Airborne microbial challenges of blow/fill/deal equipment: a case study. *Journal of Parenteral Science and Technology*, 45: 187–192, 1991.

Center for Disease Control (CDC). Follow-up on septicemias associated with contaminated intravenous fluids. *Morbidity and Mortality Weekly Reports*, 22: 115–116, 1973.

Clothier Report. *Report of the Committee Appointed to Inquire into the Circumstances, Including the Production, Which Led to the Use of Contaminated Infusion Fluids in the Devonport Section of Plymouth General Hospital* (CM Clothier, Chairman). London: Her Majesty's Stationery Office, 1972.

Desai, K.B. Particulate matter in injectable fluids. *Eastern Pharmacist*, July 1987: 43–44, 1987.

Dougherty, S.H. Pathology of infection in prosthetic devices. *Reviews of Infectious Diseases*, 10: 1102–1117, 1988.

Elek, S.D., Conen, P.E. The virulence of *Staphylococcus pyogenes* for man: a study of the problems of wound infection. *British Journal of Experimental Pathology*, 38: 573–586, 1957.

Felts, S.K., Schaffner, W., Melly, A., Koenig, M.G. Sepsis caused by contaminated intravenous fluids — epidemiological, clinical and laboratory investigation of an outbreak in one hospital. *Annals of Internal Medicine*, 77: 881–890, 1972.

Fernandez, C., Wilhelmi, I., Andradas, E. et al. Nosocomial outbreak of *Burckholderia pickettii* infection due to a manufactured intravenous product used in three hospitals. *Clinical Infectious Diseases*, 22: 1092–1095, 1996.

Fredrickson, A.G. Stochastic models for sterilisation. *Biotechnology and Bioengineering*, 8: 167–182, 1966.

Hall, L.B. NASA requirements for the sterilization of spacecraft. In *Spacecraft Sterilization Technology*. Washington, D.C.: NASA 1965.

Hall, L.B., Lyle, R.G. Foundations of planetary quarantine. In *Planetary Quarantine, Principles, Methods and Problems,* ed. Hall, L.B. New York: Gordon and Breach, 1971.

Kallings, L.O., Ringertz, O., Silverstolpe, L., Ernerfeldt, F. Microbial contamination of medical preparations. *Acta Pharmaceutica Suecica*, 3: 219–228, 1966.

MacIntosh, C.A., Lidwell, O.M., Towers, A.G., Marples, R.R. The dimensions of skin fragments dispersed into air during activity. *Journal of Hygiene*, 81: 471, 1978.

Maki, D.G., Rhame, F.S., Mackel, D.C., Bennet, J.V. Nationwide epidemic of septicemia caused by contaminated intravenous products. *American Journal of Medicine*, 60: 471–485, 1976.

NASA Standard Procedure for Microbiological Examination of Space Hardware. Document no. NHB 5340.1.A. (Available from) Washington, D.C.: Superintendent of Documents, U.S. Government Printing Office, 1968.

Noble, W.C., Lidwell, O.M., Kingston, D. The size distribution of airborne particles carrying microorganisms. *Journal of Hygiene*, 66: 385, 1963.

Roark, A.L. A stochastic bioburden model for spacecraft sterilization. *Space Life Sciences*, 3: 239–253, 1972.

Sykes G. The control of airborne contamination in sterile areas. In *Aerobiology — Proceedings of the 3rd International Symposium*, ed. Silver, I.H. London: Academic Press, 1970.

Wagner, C.M., Akers, J.E. *Isolator Technology — Applications in the Pharmaceutical and Biotechnology Industries*. Buffalo Grove, Illinois: Interpharm Press, 1995.

Whyte, W. Settling and interaction of particles into containers in manufacturing pharmacies. *Journal of Parenteral Science and Technology*, 35: 255–261, 1981.

Whyte, W. The influence of clean room design on product contamination. *Journal of Parenteral Science and Technology*, 38: 103–108, 1984.

Whyte, W. Sterility assurance and models for assessing airborne bacterial contamination. *Journal of Parenteral Science and Technology,* 40: 186–197, 1986.

Whyte, W. *Cleanroom Design*. Chichester, U.K.: John Wiley & Sons, 1991.

Whyte, W., Bailey, P.V. Reduction of microbial dispersion by clothing. *Journal of Parenteral Science and Technology*, 39: 51–60, 1985.

Whyte, W., Bailey, P.V., Tinkler, J., McCubbin, I., Young, L., Jess, J. An evaluation of the routes of bacterial contamination occurring during aseptic pharmaceutical manufacturing. *Journal of Parenteral Science and Technology*, 36: 102–107, 1982.

Whyte, W., Niven, L., Bell, N.D.S. Microbial growth in small-volume parenterals. *Journal of Parenteral Science and Technology*, 43: 208–212, 1989.

Chapter 2

Microbiological Environmental Monitoring

Nigel Halls

CONTENTS

1 INTRODUCTION

The two main expressions used in relation to the operation of pharmaceutical clean rooms are not synonymous: environmental control and environmental monitoring.

0-849-32300-2/04/$0.00+$1.50

Environmental control describes the systems functionally ensuring that clean rooms operate within predetermined limits. There are many such systems, integrated and overlapping, described in Chapter 3.

Environmental monitoring describes the techniques used to measure the effectiveness of the environmental control systems, and defines the procedures necessary in the event of limits being exceeded.

Environmental control, particularly in sterile manufacture, is achieved by means of many factors: well-designed and efficiently operated facilities and air-handling systems, by the use of integral HEPA filters, well-designed and well-made garments, by reliable disinfection regimes, and by rigid adherence to aseptic disciplines.

Information on the operation of all such factors is obtained from a variety of physical monitors. These include pressure differentials, air flows, supervision and other systems that can, if required, be linked to feedback control. Pressure differentials *may be lost for short periods* with minimal impact on sterility assurance, and occasional lapses in aseptic disciplines can never be totally excluded. Environmental control is best achieved by physical means, by feedback, and by automated alarms — means that respond in "real time" and can lead to immediate correction of lapses.

Microbiological environmental monitoring, however, looks indirectly at the environmental control systems. It is intended to measure the end product of such systems, i.e., the microbiological quality of the clean room.

Microbiological environmental monitoring has no immediacy. Results are not obtainable until days after the data collection, and later than the events that the data describe were occurring. Rarely are adverse microbiological results reproducible on re-examination. So often they are a case of too little, too late.

From experience, microbiological environmental monitoring is a necessary and valuable means of disclosing lapses in control, which may not be signalled by any other means. This most typically happens with regard to personnel. The periodic presence of a quality assurance (QA) microbiologist taking environmental samples is undoubtedly a reminder to production personnel of the importance of asepsis, particularly if the environmental microbiologist is also involved in aseptic training and periodic retraining of the operators. The microbiologist's own practices and techniques must be beyond reproach.

Conversely, all time-served production operators remember occasions when they may have "screwed up" only a few minutes before a microbiological monitoring, and escaped with satisfactory results. They will also know of many occasions when they could, "hand on heart" with absolute certainty, testify that they had done nothing wrong, but the microbiological results indicate the contrary. In other words, the results of microbiological monitoring are erratic.

Environmental monitoring is one of the frequently criticised areas in regulatory inspections. An FDA inspector once said to the author: "Finding problems with environmental monitoring techniques, programs, results and responses is like shooting fish in a barrel!"

Good environmental microbiologists understand sound laboratory controls and have detailed facility and process knowledge. Good microbiological monitoring programs are unique to particular facilities; they must focus both on known vulnerabilities (validation may disclose these) and on discovering unknown vulnerabilities. Good environmental microbiologists analyse their data regularly, looking for changes to established patterns and for trends.

The purpose of microbiological environmental monitoring is to discover the unexpected, unpredicted vulnerability of facility or process to microbiological contamination. Its limitations are speed of response (too slow) and consistency (erratic nonreproducible results).

2 ENVIRONMENTAL MONITORING: APPLICATIONS AND LIMITS

All pharmaceutical manufacturing environments merit a level of environmental monitoring. The greatest emphasis and the tightest limits are applied to sterile manufacturing facilities. When different areas within sterile manufacturing facilities serve different purposes, so the environmental monitoring programs differ. The question being so often asked is: What limits should be applied in microbiological monitoring of sterile products manufacturing facilities?

In Europe the answer is easy. Microbiological limits applying to various grades of manufacturing clean room are specified in the 2002 *Guide to Good Manufacturing Practice for Medicinal Products* (MCA, 2002). Monitoring should be done when the facilities are manned and operational. Table 2.1 summarizes the main microbiological limits taken from this document.

In the U.S., the *United States Pharmacopeia* (*USP*) has a general Chapter <1116> on the topic of microbiological environmental monitoring. The limits are broadly (within the variability of microbiological technique) the same as those of the European community. There has been some controversy in the U.S. over the need for this Chapter.

USP contends that Chapter <1116> fulfills a "customer requirement" for guidance on how much microbiological contamination is tolerable in aseptic manufacture. The notion of there being a customer requirement appears to be supported by some 40 to 70% of the respondents to the 1997 Parenteral Drug Association (PDA) survey, who claimed that at least some of their environmental microbiological limits were based on guidance from regulatory or compendial bodies.

Since the limits are contained in a nonmandatory General Chapter, *USP* believes they cannot logically be perceived as overly restrictive.

Akers (1997) expresses contrary arguments. Irrespective of the *USP*'s stance on these limits being nonmandatory, they will be perceived by the pharmaceutical industry and enforced by the regulatory agencies as if they were mandatory, and that limits of this type will not necessarily serve the greater good of pharmaceutical manufacture.

Table 2.1 European Union (EU) Recommended Upper Limits (Abbreviated) for Microbiological Environmental Monitoring of Clean Rooms

	Active air sample (cfu/m^3)	*Settle plate, 90 mm (cfu/4-hour exposure)*	*Contact plate, 55 mm (cfu/plate)*	*Glove print, five fingers (cfu/glove)*
Grade A (local zones for high-risk operation, e.g., point-of-fill, protection of aseptic connections, etc.)	<1	<1	<1	<1
Grade B (e.g., in the case of aseptic manufacture Grade B is the background environment for Grade A zones)	10	5	5	5
Grade C (e.g., rooms where aseptic solutions are prepared for filtration)	100	50	25	NA

Abbreviations: cfu, colony forming units; NA, not applicable.

Akers effectively argues that environmental data tracking (qualitative as well as quantitative) in ways that can lead to the recognition of changed or changing circumstances, is more effective in detecting loss of control than rigid adherence to standard limits that may have no bearing on the actual condition of the facility being monitored.

Both positions are of value.

Published guidelines provide benchmark environmental targets for new facilities, and may help reduce the apparent subjectivity of demands for action from microbiologists and QA specialists when they perceive that environmental conditions have deteriorated significantly.

However, they must be sensibly supplemented by additional limits, set around actual environmental performance of a facility, so that sudden or progressive deteriorative changes in the actual condition elicit prompt corrective responses, irrespective of whether the published limits have been breached. *Local action limits* (qualitative as well as quantitative) must be established. These limits should usually operate at tighter levels than the published limits, and should *never* be allowed to be weaker. Paradoxically, it is possible that two facilities operated by the same company, but built to different design or fabric standards, or operated by different groups of personnel, could require different environmental action limits.

Quantitative microbiological limits must be seen for what they are — crude measures of environmental stability, and so must be interpreted intelligently. The E.U. limits given in Table 2.1 are addressed here not as a criticism of the limits themselves, but as an illustration of some of the aspects of limits and limit setting. These must be treated with caution. A great deal of energy, time and money is expended in trying to harmonize such limits with respect to distinctions of no real long-term consequence to environmental control.

Table 2.1 shows a whole series of limits of less than one colony-forming unit (cfu). Limits of less than 1 cfu, for instance, must mean that 0 cfu per plate is the only acceptable result for contact plates and glove prints. This assumes no incidental media contamination, in preparation or in handling. Microbiologists are imperfect — some frequency of incidental media contamination is inevitable; it is usually low but it cannot be ruled out. Preincubation of media is no guarantee against incidental contamination while the test is carried out.

The E.U. limit of less than 1 cfu per four-hour settle plate also falls into this category. In theory one might expose a plate for longer than four hours to improve the sensitivity of the method but, in practice, four hours is near the outer limit of drying out with consequent loss of growth supportiveness. The limit of less than 1 cfu per m^3 set for active sampling of Grade A conditions either means 0 cfu per m^3 sample size or forces the microbiologist to use a larger sample size; e.g., a 2-m^3 sample size allows for 1 cfu as incidental contamination, a 3-m^3 sample size allows for 2 cfu, and so on.

How much wiser it would have been had all of these limits been set at *no more than* 1 cfu. It would have made no difference to environmental control, but would have allowed for some incidental contamination and reduced the frequency of false responses.

In Table 2.1 the distinction between glove prints between Grade A (less than 1 cfu) and Grade B (no more than 5 cfu) conditions is extremely interesting. Its intent is to ensure that personnel who have to make truly aseptic adjustments in areas where product is exposed (e.g., in or close to the filling zone) have the more severe restriction placed upon them commensurate with the seriousness of the work they have to perform.

Personnel required to do heavier, cruder work, such as loading or unloading autoclaves are, according to these limits, allowed up to 5 cfu per glove print. In practice, no responsible aseptic products manufacturer should ever allow personnel who work in filling rooms, whatever their task responsibilities, to persistently return glove-print data of 2 to 5 cfu.

Personnel are, in well-designed facilities with good air-handling systems, the greatest risk to asepsis. They require training, are required to disinfect their hands frequently either in disinfectant dip bowls or by spraying. In the author's experience the typical data pattern from good operators in well-managed facilities is 0 cfu per glove print, with the occasional 1 cfu emerging as an indicator of the technique's limitations. The E.U. limit for Grade B areas is arguably dangeriously misleading.

3 ENVIRONMENTAL MONITORING: METHODS

The four sets of microbiological limits in Table 2.1 point to the four main microbiological environmental monitoring methods:

- active air sampling
- passive air sampling (settle plates)
- surface sampling
- personnel sampling

The European enthusiasm for settle plates is not reflected in U.S. practice. Conversely, contact plates have not been taken up as eagerly in the E.U. as by the U.S. Both techniques have their limitations. The technique of surface swabbing should be added to these four. Personnel monitoring may not be restricted to glove prints.

3.1 Active Air Sampling

Active air sampling is intended to provide an index of the number of microorganisms per unit volume of air space in clean rooms. If the clean room is served by a good air-handling system, with integral HEPA filters in place, airborne microbial contamination arises from personnel operating within the clean room.

Most active air sampling should be done when the clean room is operational. However, if a clean room has been nonoperational for a few days (e.g., a long weekend) or a few weeks (e.g., a scheduled shutdown for vacation or maintenance), it is beneficial to start sampling a few days prior to production start-up.

All active air samplers will disrupt air flow to some extent. They should be located carefully, and when they are operated (none work on the continuous sampling principle), they must not counteract protective air flow patterns in significant parts of the clean room.

Active air sampling is the only microbiological environmental monitoring process involving serious capital expenditure. Active air samplers cost several thousands of dollars, insignificant compared with the costs of production autoclaves, aseptic filling machines, etc., but significant enough in terms of the cost of laboratory equipment for there to be lively competition between suppliers.

It is important to understand the distinctions between active air samplers. Figure 2.1 summarizes some of the main characteristics of available active air samplers.

1. The "traditional" active air sampler is the slit-to-agar sampler of the Andersen sampler, Casella sampler, Mattson Garvin sampler, etc. types as shown diagrammatically in Figure 2.1. These have become the "standard" against which other samplers are compared.

Slit-to-agar samplers of this type have been widely used for monitoring pharmaceutical clean rooms and have in fact dominated the market for many years.

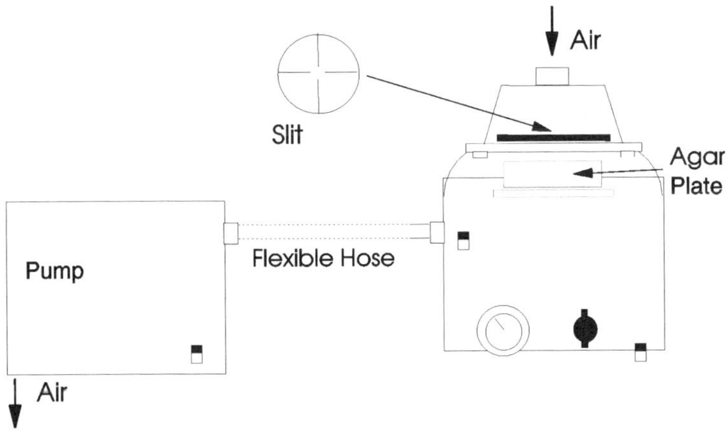

Figure 2.1. Simplified representation of a slit-to-agar sampler.

They are heavy, thus limiting their portability, even though they are usually mounted on trolleys or carts to provide some flexibility, and they can be fumigated but are difficult to disinfect. They are best dedicated to one clean room and maintained "captive."

The Petri dishes used in slit-to agar samplers are of a nonstandard size (150 mm) requiring a fill volume of about 100 ml. They are unsuited to most automated plate pourers. The cost of media is increased at least five-fold over samplers that use standard size Petri or Rodac Petri dishes.

2. A second type of active sampler is the Reuter Centrifugal Sampler (RCS). Two types of instrument are marketed, the RCS and the RCS-Plus. Both are battery-powered, lightweight, portable and easy to fumigate and disinfect.

In the first development of the RCS (still marketed and used widely), air is drawn into an open-fronted cylindrical housing by means of a low-pitch impeller (Figure 2.2). The air drawn into the housing is redirected back out again through a turn of about 360°. Some of the air is forced towards the inner wall of the housing, where an agar-containing flexible plastic strip is located around the circumference. The cone of air deflected forward from the RCS sampler interferes with protective air-flow patterns in areas where it has been used (Kaye, 1988).

In its later development, the RCS-Plus, the air is exhausted through ports at the rear of the impeller head, successfully reducing the air-flow interference effect (Ljungqvist and Reinmuller, 1991).

The earlier of the two developed instruments, the RCS, has a stated air-intake rate of 280 litres per minute of which, the manufacturers claim, about one-seventh (40 litres of air per minute) is directed onto the agar strip. There has been much debate

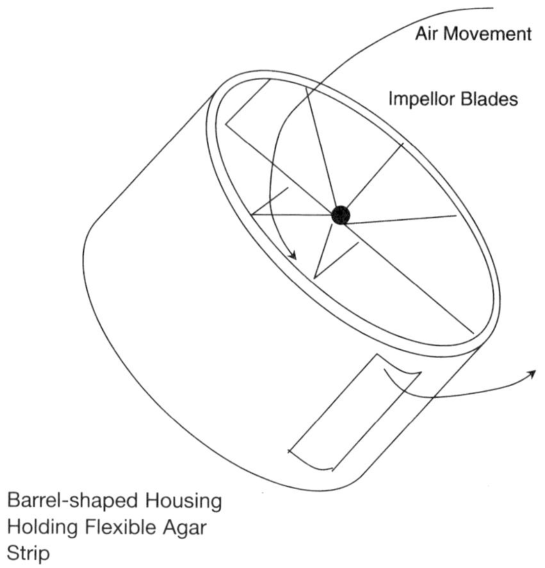

Figure 2.2. Simplified representation of an impeller head of an RCS active air sampler.

as to whether this instrument or its successor is capable of collecting all particle sizes that might carry airborne microorganisms, if it even collects particles of the sizes most likely to carry airborne microorganisms, and if it can be regarded as a truly quantitative instrument for measurement of airborne microorganisms against published limits.

It seems that the RCS centrifugal sampler is less efficient at capturing airborne microorganisms than other types of available samplers. The question though is: Does it really matter?

The answer: It depends on the circumstances. The RCS effectively samples at 40 litres per minute for a maximum sampling time of five minutes, to a maximum sample volume of 400 litres. It is unsuited to sampling E.U. Grade A environments (laminar flow-protected aseptic areas) because, even if there were no controversy over other aspects of this sampler, it quite simply does not draw a large enough sample to verify compliance with the limit of less than 1 cfu per m^3 (1000 litres).

Its limitations are insignificant in Grade C environments against a limit of no more than 100 cfu per m^3 (preparation areas for aseptic manufacture or filling rooms for terminally sterilized products), but borderline for Grade B environments (aseptic filling rooms). It is probably wisest to confine this instrument to Grade C environments for sterile manufacture, and to sampling nonsterile manufacturing environments.

The RCS-Plus instrument samples at 50 litres per minute and can be operated for 20 minutes to collect a 1-m^3 air sample. There are data demonstrating that its efficiency of collection of microorganisms on particles of 4 μm with which most microorganisms are believed to be associated, and larger, is comparable to slit-to-agar samplers (Benbough, 1992). This instrument is suitable for sampling all grades of pharmaceutical clean room, is less robust than the RCS, and requires frequent calibration of its sampling heads.

Centrifugal samplers are generally in the same price range as other types of samplers but running costs are higher because of the need to purchase the agar-containing flexible strips. Empty strips can be purchased for local laboratory filling but involve introducing a nonstandard operation into laboratories.

3. Filtration is the third means of active air sampling, but death of microorganisms by desiccation on the membranes has restricted its application. The most widely used variant on filtration, the Sartorius MD-8 sampler, avoids desiccation by using gelatin membranes. After sampling, the gelatin membrane is transferred to an agar plate where it dissolves and merges into the agar during incubation (plates should be well dried before transfer to prevent colonies from spreading). Efficiencies of collection and recovery of microorganisms are comparable to slit-to-agar samplers (Parks et al., 1996; Pendlebury and Pickard, 1997).

The MD-8 is battery-powered, heavier and bulkier (because of the pump system) than the centrifugal samplers but is still portable and easy to fumigate and disinfect.

The maximum sampling speed of the MD-8 is about 130 litres per minute. Flow rates are adjustable to match the local airflow and enable isokinetic sampling. This makes the MD-8 the most useful of all the samplers for evaluating the airborne microbial counts in laminar flow-protected areas (e.g., at point-of-fill) while operational. All other active air samplers are required to be used with considerable care in these areas, to avoid the risk of disrupting protective air flow. This is greater than the benefit of having data to show that microorganisms have been excluded.

4. At least three variants of another type of air sampler exist, less commonly used than in monitoring pharmaceutical clean rooms. These operate on the principle that air is drawn through a perforated atrium head where air samples are collected by impaction on an agar plate. The air is exhausted at the rear (beneath the agar plate). The location of the exhaust forces the sampled air, after its initial impaction, to turn through 90° and flow over the surface of the agar to exhaust. This may help to improve microbiological collection efficiency.

The heaviest of the three models is comparable to the MD-8 with respect to portability and ease of fumigation and disinfection. Models are marketed by PBI International, Merck and by Veltek Associates (VAI) (Figure 2.3).

These samplers are operated by constant-speed pumps, and sample volumes are controllable through a timer setting.

The agar used in the Merck and VAI samplers is poured in standard 90-mm Petri dishes and for the PBI sampler in Rodac (55-mm) Petri dishes. Their running costs are far less than those of the other sampler types.

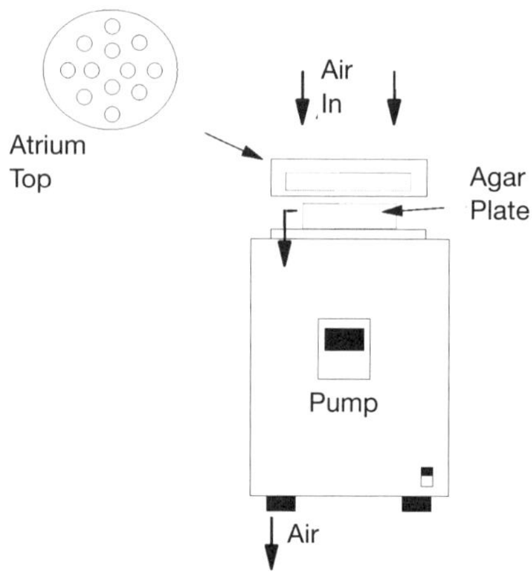

Figure 2.3. Simplified representation of an active air sampler with perforated atrium head.

Table 2.2 summarizes some of the characteristics of available active air samplers.

3.2 Passive Air Sampling

Passive air sampling is done by means of settle plates: agar plates are left open and exposed in clean rooms for defined periods. They are used widely in Europe where they have been strongly advocated over many years in the work of Whyte (1986), but less so in the U.S., except in facilities manufacturing for export to Europe.

The principles of the settle plate were empirically demonstrated by Whyte (1986). Most airborne microorganisms are associated with physical particles of 12-µm diameter or larger (Whyte, 1986), which are heavy enough to settle out of air by gravity.

Sykes (1970), in earlier research, challenged this concept. He calculated a mean settling time of seven minutes for particles of 5-µm diameter (the particle size normally associated in European regulatory literature with airborne microorganisms) through a one-foot column of still air. Probably only the heaviest particles will be collected on settle plates laid out horizontally (as is the almost universal practice) on flat surfaces in turbulent or unidirectional air flow clean rooms.

It is also important to consider the significance of 1 cfu on a settle plate. What does it represent? One single viable microorganism? Or several tens or hundreds of microorganisms carried on a single skin particle? This restricts the value of settle plates for consistent comparison with quantitative limits, though their qualitative value is not debatable.

Table 2.2 Some Characteristics of Available Active Air Samplers

Sampler	Sampling rate (litres/minute)	Weight (kg)	Recovery of microorganisms
Slit-to-agar sampler (Casella)	175, 350, 525 or 700	16 (with pump)	Agar in 150-mm Petri dishes
Slit-to-agar sampler (Mattson Garvin)	35	~ 7	Agar in 150-mm Petri dishes
Centrifugal sampler (RCS)	40 (effective)		Agar in flexible plastic strips unique to centrifugal samplers
Centrifugal sampler (RCS-Plus)	50 (effective)	~ 1.5	Agar in flexible plastic strips unique to centrifugal samplers
Filtration sampler (Sartorius MD-8)	130	–	Gelatin membrane transferred to standard 90-mm Petri dish containing agar
Perforated atrium sampler (PBI)	90	~2	Agar in 55-mm Rodac Petri dishes
Perforated atrium sampler (Merck)	100	~ 2	Agar in standard 90-mm Petri dishes
Perforated atrium sampler (VAI)	175	~ 5	Agar in standard 90-mm Petri dishes

There has been some debate about how long settle plates may be left open in clean rooms before the effects of desiccation impair the ability of the agar to support the growth of microorganisms. The PDA recommends 30 minutes in its 1981 Monograph No. 2 (PDA, 1981), the Parenteral Society in the U.K. recommended four hours in 1990 (Parenteral Society, 1990), while Whyte and Niven (1987) argued that the viability of microorganisms on agar plates was not significantly affected by desiccation for exposure periods of up to 24 hours. It is most likely a function of the depth of agar in the Petri dish and the condition of the agar when introduced into the clean room.

All practising microbiologists will have some experience of seeing agars drying out, and agars with desiccated "skins" on their surfaces. They should have technical or procedural mechanisms in their laboratories to prevent such plates from being used, or at least to ensure that data from such plates are discarded.

Regardless of these objections settle plates are popular with many microbiologists, as they are inexpensive, do not disrupt, in most instances, protective air flow patterns, and require no great technical expertise to generate data. They can sometimes be laid out by production rather than QA personnel, though some regulators have insisted that these practices should be observable for independent QA audit.

3.3 Surface Sampling

There are two methods available for sampling surfaces: the contact plate and the swab. Only the contact plate is referenced in the E.U. requirements and in *USP* Chapter <1116> limits. Swabs are not thought, generally speaking, to produce quantitative data.

In the contact plate, agar contained in a specially designed Petri dish, the Rodac plate, is pressed against the surface being sampled. Microorganisms are transferred from the surface to the agar and colonies develop on incubation. Devices are available to ensure that Rodac plates are applied with a controlled, consistent, even pressure.

In the pharmaceutical industry, practically every surface in clean rooms is sampled by Rodac plates at some point. However, these plates are really only suited to sampling flat surfaces. Paradoxically flat surfaces are the easiest to clean and disinfect, and are also the least likely locations for persistent microbial colonization.

Rodac plates should not be used in critical areas such as machine surfaces around point-of-fill, on hoppers, etc., while manufacture is in progress. The risks of contamination to the product from the test procedure are greater than the benefit afforded by the data. They are usually routinely tested at the end of a filling batch or campaign.

Consideration must be given to cleaning up any residual nutrients that may have been transferred from the agar to the sampled surface. The difficulty in doing this is often exaggerated by opponents of the use of the Rodac plate. Neutralizers may have to be incorporated in the agar if it is used on disinfected surfaces.

Swabbing is done by scrubbing a moistened cotton, nylon or alginate bud across a nominal surface area. The bud is then either rolled across the surface of an agar plate, or agitated in a known volume of sterile water or other noninimical but nongrowth-supporting milieu, from which a sample is taken and plated on agar. Alternatively, the swab may be broken off into a tube of "enrichment medium," which is then incubated, and any growth streaked on agar.

As with contact plates swabs share the problems of invasiveness (the potential to contaminate a previously uncontaminated surface and thence to contaminate the product), and of the impairment of growth support of their media by disinfectant traces. Conversely, they are not restricted to flat surfaces, but are ideal for examining crevices, niches and concealed and roughened surfaces — the most difficult to clean and disinfect.

In the quest for quantitative data it is not unusual for swabs to be used in conjunction with templates defining a particular surface area to be swabbed. Regrettably this practice detracts from the best use of swabs, in the areas where templates and Rodac plates cannot be used. Swabs generate qualitative data, semi-quantitative at best, nonetheless their value cannot be underestimated.

Surface sampling has its greatest value on commissioning a new clean room, or restarting an existing clean room after a shutdown period, where it has been allowed to "go nonsterile." After an initial cleanup, surfaces should be sampled; the clean room should then be disinfected and sampled again. The clean room should be thoroughly disinfected (second disinfection) and sampled a third time, and quarantined until the results from the first and second samplings are available.

If these data are satisfactory, the clean room may be released to production. If not, the room should be disinfected (third disinfection) and sampled (fourth sampling) again, but may be released to production if the data from the third sampling are satisfactory.

In routine use, surface sampling in critical areas should be done at the end of batches or campaigns. Surface samples in less critical areas may be done while sterile manufacture is in progress.

3.4 Personnel Monitoring

Personnel are probably the greatest source of microbial contamination in clean rooms. People are mobile, unpredictable and cannot be sterilized (in the microbiological sense of sterilization). Some may be greater potential sources of contamination than others; there may be sudden or periodic changes in their contamination potential for physical or even psychological reasons — the garments provided to them may be inappropriate (wrong fabric or fit); and none of this may be obviously evident to supervisors. It makes good sense to monitor personnel for microbiological contamination.

The main type of personnel monitor is the glove print (also referred to as a finger print, finger dab, etc.) where the tips of four fingers and the thumb of each gloved hand are pressed on an agar plate. This can be done at any time within a production shift; personnel should be taken into the changing room to make the glove print, and the gloves should then be discarded and replaced. The argument can be made that replacement of gloves after glove prints is unnecessary, because disinfection of gloved hands has been validated. The cost of a few extra pairs of gloves is insignificant in comparison to the cost of a batch of product rejected, or a patient harmed.

Other less common approaches to personnel monitoring include swabs or contact plates from garment surfaces. None of these should be allowed except at the end of a shift, when the garments are due to be discarded. The barrier properties of garment fabrics are impaired when wetted, as they are by swabbing.

Practically every part of the front of garments is monitored somewhere, most frequently the forearm, for obvious reasons. The chest region in front of the armpit gusset is often sampled because of the strain on the garment in movement, and the possible effects of perspiration. Masks and head covers are sometimes examined, particularly for operators who have difficulty in avoiding touching their faces. Swabs on overboots are also sometimes taken, but allowances must be made for periodic high counts.

4 ENVIRONMENTAL MONITORING: MICROBIOLOGICAL CONSIDERATIONS AND CONTROLS

Tryptone Soy Agar (TSA) is the standard medium for microbial recovery in environmental monitoring programs. Incubation is at 30–35°C. Yeasts and moulds may also be specifically sought out. Sabouraud Agar (SA) incubated at 20–25°C is generally used for this purpose. Media used in antibiotic manufacturing facilities, particularly when they are solid dosage forms, should be reviewed for their capability to recover environmental microorganisms. Neutralizers may have to be included in the preparation of the media. β-lactamase is included in media used in environmental monitoring programs for facilities manufacturing penicillins and cephalosporins.

The duration of incubation generally recommended is 48 to 72 hours. This is curious because it means that samples taken on Thursdays (as they surely must in at least some weeks) must be read on Saturday or Sunday, when most laboratories are not routinely staffed. Since the upper time limit on incubation is largely arbitrary, it makes more sense to specify incubation to be 48 to 96 hours.

The occurrence of spreading forms that can obscure other growths after lengthy incubation on agar is fairly unusual in pharmaceutical manufacturing environments.

There is a regulatory enthusiasm for manufacturers to include some consideration of anaerobic microorganisms in environmental monitoring programs, usually by incubating TSA in anaerobic conditions. Other media more suited to the recovery of anaerobes may be used.

Anaerobic environmental monitoring may be seen as fulfilment of a regulatory requirement. Obligate anaerobes are intrinsically unlikely to be present in most pharmaceutical manufacturing environments; oxygen is toxic to obligate anaerobes. Pharmaceutical clean rooms are continuously swept by filtered air, and surfaces are smooth and clean; consequently there should be few opportunities for anaerobes to survive, even fewer for them to be recovered by active or passive air sampling, or on contact plates.

In practical terms, pharmaceutical clean rooms should be "brainstormed" to determine any locations where anaerobic microorganisms may survive, e.g., where there is grease, or in oil sumps. These locations should preferably be engineered out of the clean room, but if this is impossible, they should be the focus of anaerobic environmental monitoring by swabbing.

Conversely, microaerophilic organisms (e.g., *Propionibacterium* spp.) are rarely isolated in routine environmental programs but are not unknown as sterility test contaminants on those rare occasions when sterility tests "fail." Swab enrichment in fluid thioglycollate medium enables an evaluation of the presence of these common human commensals in the manufacturing environment.

The media used for environmental monitoring should be demonstrably able to support the growth of microorganisms (pharmacopoeial types and local environmental isolates as indicated for media fills). Although it is not absolutely necessary that growth support checks should always be carried out on every autoclaved batch of media, every supplier's batch of dehydrated media should certainly be checked. Some FDA investigators insist that every autoclaved batch of environmental monitoring media is tested for growth support.

Environmental control media should be validated for their ability to support growth throughout their shelf lives. Agars are often prepared, sterilized and stored for days or weeks before melting and pouring as environmental monitoring plates.

Environmental media must be preincubated for sterility before they are taken into Grade A and Grade B aseptic areas. There are two reasons for this.

First, it is quite absurd, when faced with the very strict microbiological limits usually applied to these areas, to risk reacting to incidental contamination arising from media preparation.

Second, aseptic areas must not be compromised by taking contaminated agars into them. This means that considerable care should be taken to ensure that environmental plates are not cross-contaminated in microbiologically contaminated incubators. Plates should be poured and double-bagged in plastic in Grade A laminar air flow-protected areas, before preincubation. Where agar plates have to be dried, this should be done in the Grade A pouring areas prior to bagging.

5 ENVIRONMENTAL MONITORING: ESTABLISHING A PROGRAM

Regulatory agencies require that environmental monitoring programs for sterile manufacturing facilities be defined and documented, with respect to where and how often samples should be taken. In Europe, proposals were put forward by the Parenteral Society in 1990 (Parenteral Society, 1990). In the U.S., Agalloco (1996) published a table of "possible sample frequency for routine monitoring." Both sets of general recommendations implicitly consider facilities in frequent or constant use. Neither makes any specific reference to those facilities in which sterile products are manufactured infrequently, or on a campaign basis.

Both sources may be used as guidance to establishing an environmental monitoring program, but they cannot be seen as definitive. Each sterile manufacturing facility is unique — in technology, manning, design, and in use. The following principles are valuable in designing specific environmental monitoring programs.

1. The criticality of particular areas must be taken into account when determining their environmental monitoring frequencies.

- Grade C areas (Table 2.1) are not intended to be aseptic; personnel are generally not required to wear sterile garments; nonsterile vessels are in use for bulk compounding. The reason for their microbiological control is protection and minimization of the microbiological challenge to the aseptic areas (Grades B and A of Table 2.1). There is little immediate need for microbiological environmental monitoring data in Grade C areas. There should, however, be sufficient data to indicate that these measures remain under control when the areas are operational.

Monitoring can be done in these areas without significant risk to the process; however, the value of the data is limited.

Weekly, fortnightly or monthly sampling would be reasonable for facilities in frequent or constant operation, depending on their history of compliance with limits and the variability seen in the measures. More variability in the measures from these areas than in aseptic areas can be expected. Where a facility is used, for example, only four times a year, it is clearly unrealistic to monitor weekly or monthly when there is no activity. Such facilities should be monitored only when operational, when monitoring is done at a greater frequency than more heavily used facilities. There would be only limited value in obtaining, for example, only one Grade C datum point for each measure in a facility that is only used for one month at quarterly intervals.

- The intermediate area between Grade C areas and aseptic filling rooms are the change rooms (often called "white" change rooms). These are required to meet the environmental standards of the area to which they give access (MCA, 2002). However they are exposed to a greater microbiological challenge as a result of personnel entering in nonsterile garments, stripping off and changing — in other words there is more exposure of the change rooms to microbiological contamination, and possibly a greater level of physical activity to disseminate contaminants. The risk is that the change rooms themselves become a source of contamination, which leads to contamination of the aseptic rooms with the operators as the vectors. Personnel are always trained in changing disciplines but they are never, for clear unarguable personal reasons, routinely supervised. Microbiological environmental monitoring of these areas is critical.

  Monitoring can be done in these areas without significant risk to the process — a major value.

  Passive air sampling by settle plates is a good means of evaluating the changing process. The plates may be placed on the floor, on stepover benches, and on table tops over the time that personnel are stripping and changing; there is no value in laying out settle plates in empty change rooms. Surface sampling is also valuable; Rodac plates or swabs are equally applicable. Surface sampling should be concentrated on stepover benches, table tops, walls, in the "clean" side of the rooms after personnel have passed through. Personnel should not touch and contaminate these surfaces. However often the conditions in change rooms are so cramped and poorly designed that surface contact is unavoidable. Managers rarely enter aseptic areas and when they do they most often associate the difficulties they experience with their own awkwardness rather than with the facilities provided. Floor sampling, although supported by the FDA, is relatively meaningless. Passive air and surface sampling should be done in "white" change rooms on every operational day or shift. Active air sampling may give unreliable results in change rooms while personnel are changing. The samplers may upset protective air patterns and there may be insufficient total room-air volume to dilute contamination to a fairly representative level. Active air sampling should be done in change rooms when personnel have left (either to work in the aseptic facility or at the end of the working day).

- Grade A and Grade B aseptic areas should not be microbiologically contaminated. Environmental monitoring data from these areas are likely to be most valuable in diagnosing sterility test or media fill anomalies, and to improving environmental controls. The value of obtaining microbiological monitoring data from these areas is significant, but the risk of contaminating the areas, process or product in obtaining these data is equally important
- The three most valuable measures in these areas are active air sampling, surface sampling by swabbing, and personnel monitoring. The number of air changes in modern, well-designed aseptic filling rooms, and particularly in laminar air flow-protected areas, is likely to counteract the gravitational forces necessary for microorganisms to fall out on settle plates. The extensive use of disinfectants in Grades A and B aseptic areas and the comparative ease of cleaning smooth flat surfaces minimizes the value of the Rodac plate. Data from active air sampling, surface sampling and personnel monitoring should be obtained for every shift or for every batch of product manufactured and at regular, frequent intervals. In a facility operates daily, batch-by-batch data will serve as periodic data. In a facility that is only occasionally used, environmental data from Grade A and Grade B areas should be obtained at daily intervals for at least three days before scheduled production start-up, so that some microbiological measures have completed incubation before routine production begins. The availability of such data may in some facilities be proceduralized as a requirement for allowing production start-up.

2. The activities necessary for obtaining environmental monitoring data must never be allowed to compromise the processes of environmental protection. This is most important in Grade A and Grade B aseptic areas.

- All forms of environmental sampling, except active air sampling, can be done in "background" aseptic filling rooms (Grade B) without significant risk to process or product. Only in the most compact filling rooms is the presence of a QA microbiologist for a few minutes in each shift likely to be significantly disruptive to air flows or operational disciplines. It should be remembered that the QA microbiologist is a fallible mortal from whom contamination may arise. Particular care should be taken with active air samplers that their exhaust air is not directed towards areas where product or product-contact components are exposed.

A comprehensive set of monitoring locations should be developed from:

- past experience of where contamination has actually been found
- opinion of where contamination may most likely occur
- knowledge of the locations where the highest nonviable particulate counts were obtained in validation.

All of these locations need not be monitored for every batch of product manufactured or on every monitoring occasion. A planned or randomized matrix approach may be taken. All locations should be carefully monitored over a reasonable period (e.g., weekly for a facility that is in frequent or constant operation). There should be data from each location for every day in the working week (e.g., monthly), and from each location at the beginning, the middle and the end of each working day (e.g., quarterly).

- Grade A areas are specifically protected to a high degree because the work done or the materials exposed in these areas is of particular risk. Monitoring should as far as possible avoid being intrusive. The best time to do any monitoring in these areas is at the end of the working period or shift. This is also supported by the as-yet unproven concept that microbiological contamination may, over a period of time between cleanups, build up in these areas.

Swabs should be reserved for difficult-to-clean areas. This means they are almost bound to be both intrusive and disruptive. Any traces of moisture or swab material (e.g., cotton) left behind in niches, nooks, crannies and roughened areas of filling equipment could end up damaging the product more than the value obtained from the data.

Most active air samplers will disrupt protective air-flow patterns and are therefore both intrusive and disruptive. The Sartorius MD-8 sampler can be adjusted to take an isokinetic sample and, although still intrusive, it may be used in laminar flow-protected areas with less disruption than other active air samplers.

The least intrusive and nondisruptive sampling method is the settle plate, which can only sensibly be used as a "real-time" monitor. The author's view is that the value of settle plate data does not outweigh the risk to the process of intruding into operational point-of-fill Grade A aseptic areas or into stopper hoppers, etc. This type of activity may be necessary to provide the data expected by European regulators. It is less risky to place settle plates in laminar flow units that may be used to protect autoclave off-load stations, storage cabinets, etc., and the practical value of settle plates in these areas should not be discounted. The settle plates should be set out at the time when work is done in these protected areas. Other physical monitors should be quite adequate to demonstrate that the equipment is performing as intended when personnel are not present.

As with the Grade B areas, a comprehensive list of Grade A monitoring locations should be developed. The most obvious location is at point-of-fill. Others may be on the stopper hoppers, forceps used for collecting samples or freeing jams, steriliser off-load stations, etc. Each filling room is unique and should have its own list prepared.

Locations that are only monitored at the end of the working period should preferably be monitored at the end of *every* working period. Settle plates used in laminar air-flow stations should be set out on every occasion of use as is normal in sterility testing laboratories. Operational personnel may set them out but it is probably in the best interests of "checks and balances" for microbiological QA personnel to retrieve them. Isokinetic active air samples may be monitored on a matrix basis with the same systematic or randomized coverage as recommended for Grade B areas. Other active air samples and swabs should be taken at the end of each working period but before the area and equipment are cleaned down.

Personnel working in Grade A and Grade B areas should be checked at the end of their work periods. Monitoring should be done in the change rooms. Glove prints should be done on each operator at the end of every shift. If the standards of training, disinfection, glove quality, discipline and supervision are high, the counts from glove prints are most likely to be 0 cfu, and the laboratory workload for counting and identification will therefore be low. It is best that microbiological QA supervise glove printing. Results relate back to the individual and doubtless some sensitive individuals would, if unsupervised, be tempted to take nonroutine steps to criticism, no matter how sensitively and objectively this may be offered. More complex garment monitoring may be done less frequently on a matrix basis but still covering all operators.

The program of personnel monitoring should include the personnel from microbiological QA who are responsible for environmental monitoring. Engineering and other personnel who may have reason to enter Grade A and Grade B areas should also provide, at the very least, glove prints.

3. The provision of a microbiologically controlled environment is a continuous process that is not tested and "passed" on a batch-by-batch basis.

The fact that microbiological contamination was found in a Grade A area on, for example, Tuesday when batch A was being made but not on Wednesday when batch B was being made, should not be seen as a "fail batch A, pass batch B" situation.

The situation described could equally result in "pass batch A and pass batch B" or "fail batch A and fail batch B" decisions. The discovery of contamination in microbiological environmental monitoring programs generally means that contamination was actually present (although technician-related contamination should never be discounted). The absence of contamination in a microbiological monitoring program does not mean that there was no contamination, only that none was discovered.

The presence of unexpected contamination, either with respect to numbers or types, or to the locations in which contamination may be found, should never be thought exclusive to the period in which it was detected.

6 ENVIRONMENTAL MONITORING: WHAT TO DO WITH THE DATA

The basic tenet applying to quality control data is that such data be compared to standards or limits and that decisions (usually "pass" or "fail") made on the basis of that comparison.

The underlying principle of process control data (e.g., equipment speeds, times, temperatures, pressures) is that they should be compared with predetermined limits validated as the limiting parameters of satisfactory product quality. The equipment should be then adjusted in response to the data to ensure that the process remains within these control limits.

Environmental monitoring data have a greater similarity to process control data than to quality control data, but there is little opportunity for precise process adjustment to bring controlled environments into control because microbiological indices of environmental control are crude, and limits subject to interpretation.

Nonetheless, outcomes of environmental monitoring data are the same as any other quality or process monitors. Processes may require adjustment, may have to be shut down, or the product may have to be reworked or rejected.

6.1 Responses to Infringements of Limits

Environmental standards and limits have been addressed. Manufacturers of sterile products may adopt published limits, or they may develop their own based on the performance of their facilities, or they may apply limits that are a combination of the two.This is not easy: but it is far more difficult to relate whatever limits have been decided to courses of appropriate action.

In most circumstances, environmental monitoring data should confirm that the environment is in satisfactory control. Infringements of limits should be fairly rare events. But what should be done when they do occur?

Wherever possible a two-tier approach of alert and action limits should be taken to quantitative microbiological data.

There is also a decisional element more evident in microbiology than in other quantitative sciences, as to whether a limit has in reality been breached. Where a limit has been set at, for example, no more than 5 cfu, actual counts of 6 cfu ought to be seen as infringements if the actual pattern of data is along the lines of 0 or 1 cfu, but should not be seen as an infringement if the actual pattern of data is along the lines of 4 or 5 cfu. This is a serious dilemma for QA microbiologists in the pharmaceutical industry. There is no easy answer. Education and communication with personnel responsible for general management, production, engineering and nonmicrobiological QA are probably key. However, whereas a history of reasonable flexibility ought to stand a QA microbiologist in good stead on the rare occasions when he is obliged to "draw a line in the sand," in practice he is more likely to face allegations of inconsistent application of standards.

Excursions beyond alert limits but not exceeding the action limit should initiate informal or semiformal communication with the personnel in charge of the facility or equipment, with the intention of stimulating whatever steps are necessary to avoid subsequent infringements of the action limit. Oversensitive alert limits are counterproductive.

Infringements of action limits should genuinely require action. Communication with personnel in charge of the facility or equipment must be formal. It must be in writing, elicit a written response, and the documentation must be available for inspection by the regulatory agencies. "Action" should not be interpreted as merely the "action" of documenting the infringement. Infringement of action limits must result in *meaningful* action from the personnel in charge of the facility or equipment.

Actions must be corrective (as indicated by satisfactory data from remonitoring) and should preferably be preventive. Repeated infringements of action limits, resulting in actions only along the lines of "operators were counselled and re-trained, the affected part of the facility was cleaned and disinfected" should not be tolerated. Repeated infringements of action limits indicate that the limits have been set too severely, or that the process (in its broadest sense) is not suited to manufacture of sterile pharmaceutical products.

There are a variety of processes appropriate to infringements of action limits. They are specific to particular situations. Some generalisations are, however, possible.

1. **Infringements requiring only corrective action to the process.** It is inconceivable that any operation that requires the involvement of personnel will never result in the occasional infringement of microbiological action limits. At the very least there may be occasional laboratory or microbiological QA contamination. Corrective action such as retraining and disinfection is quite appropriate as long as it is effective and the problem does not persist. One of the responsibilities of microbiological QA is to help production, engineering and cleaning personnel to decide the most appropriate course in response to action limit infringements. Responses of action limit infringements applying to Grade C areas (nonaseptic areas) need be no more severe than this.
2. **Infringements requiring preventive action to the process.** When infringement of a particular action limit persists in a particular area, or where several independent action limits are breached, there is a strong likelihood of a systematic or persistent environmental problem that must be stopped. Such problems must be investigated. It may be necessary to close the facility while

proper preventive action is implemented. Responses to infringements of action limits applying to change rooms ("white" change rooms) need not be more severe.

3. **Infringements requiring action on product.** Rarely is a sterile product rejected solely on the basis of infringements of microbiological environmental control limits. However, in relation to infringements applying to those limits in Grade A and Grade B areas, product rejection can never be totally discounted.

Information from the identity of the microorganisms isolated from Grades A and B areas may be as important as quantitative information. Action limits for these areas should include identification of all isolates. Most data from properly controlled areas of these classifications should follow the lines of 0 cfu. This recommendation should not present a significant workload. Satisfactory identification for these purposes is achievable by colonial morphology on agar and from Gram-staining. Therefore the information can be available on the same day as quantitative data.

An action limit should be set for isolation of "unusual microorganisms." This implies that there is some knowledge of the "usual" microorganisms. A responsible QA microbiologist will have a database of the usual types isolated in Grade A and Grade B areas, but all Gram-negative types should be seen as unusual enough in these areas to be perceived as an infringement of an action limit.

It is not always easy to find out what has gone wrong when an environmental action limit has been exceeded, and probably far more investigations are inconclusive than ever provide a definitive answer as to why the alert limit was infringed. Nonetheless, an infringement should be regarded as an opportunity to "test" the strengths of the sterility assurance system and to identify areas where improvement is merited. Production and engineering personnel must be involved. The identity of the microorganisms recovered can assist in focusing the investigation.

- Airborne types such as *Bacillus* spp., *Micrococcus* spp. and (to a certain extent) molds should initiate an investigation into the supply of filtered air and the maintenance of positive pressures versus less well-controlled areas. If building work is going on, attention may be given to how effectively it has been contained. The application of a sporicidal disinfectant, although recommended after all isolations of *Bacillus* spp. from Grade A and Grade B areas, may only deal with the symptom, and not the cause of the problem.
- *Staphylococcus* spp., *Propionibacterium* and some species of *Micrococcus* are usually indicative of personnel contamination. A review of past personnel monitoring data may be revealing. Frequently managers hesitate to discuss personnel monitoring data with operators unless limits are actually infringed. Sometimes an operator may be persistently giving "within-limits" counts as distinct from no recovery at all (which is the norm in Grade A and Grade B areas if hand disinfection and garment disciplines are being followed) and never be counseled until some other limit is infringed.

- Gram-negative bacteria (and to some extent molds) are water or dampness associated. Poor finishes, such as plastic-coated chipboard, can become soaked with disinfectant and yield Gram-negative recoveries.

Unfortunately every production operator knows of occasions where there have been environmental infringements when "everything went all right" and no environmental infringement when he knows there was a "screw-up." Often managerial interest in the investigative process may be sufficient to improve morale and add an edge to aseptic disciplines.

6.2 Review of Environmental Data as Part of Batch Release

All regulatory bodies expect to see environmental data reviewed as part of the batch-release procedure for sterile products. This may be confined to the data obtained from the Grades A and B aseptic areas. A summary of the data obtained during the period in which the batch was manufactured is most often included in the batch records. Alternatively, a formal statement from microbiological QA of satisfactory results may be included.

It is sensible to review environmental data in relation to batch release. However this may tend to lead to a perspective that environmental monitoring data are batch-related, rather than part of a continuous process leading to the provision of a satisfactory microbiologically controlled environment. The environmental data obtained in the periods both before and after a batch was made, may be equally as applicable as data obtained during the time the batch was manufactured. This should be appreciated in whatever mechanism is applied to a review of environmental data as part of batch release.

This review may be useful leverage to microbiological QA. Batches should not be released until environmental action limit infringement documentation has been closed out. Most batches of sterile pharmaceutical products are quarantined for seven or 14 days after manufacture, while the test for sterility is incubating. This should be sufficient time for action limit infringement reports to be released, corrective actions taken, remonitoring and the completion of documentation.

All action limit infringement reports should be sent to the batch records of the batch made at the time of the infringement and the immediately preceding and subsequent batches. Release of these batches is not permitted until the reports are formally closed out to the satisfaction of QA.

6.3 Overview of Trends in Environmental Data

Environmental monitoring data is part of a continuum. Periodic analyses of environmental data should be formally prepared, reviewed by microbiological QA, then presented for informational purposes and consideration by the production, engineering, nonmicrobiological QA, and general management personnel responsible for particular areas, operations and products.

This preparation of overview reports is in practice often done rather half-heartedly.

Excessively frequent overviews (e.g., as part of a monthly report) are as ineffective as excessively infrequent overviews (e.g., as part of annual product review where the data is collated by product rather than by manufacturing area). For heavily used facilities quarterly overviews are recommended. When facilities are used on a campaign basis the environmental monitoring data from each campaign should be overviewed separately.

The presentation of the data is extremely important. All too often this is done either as a list of action limit infringements or as pages and pages of data tabulated or plotted as counts (usually zero) against time. Neither approach is very helpful.

Good presentation is graphical. Graphical presentation offers the opportunity to personnel not closely involved in data collection to chance upon a trend, to compare one filling room with another, or to ask an informed question. There are a variety of approaches, but whatever is adopted should incorporate the following considerations.

- **Report each test separately.** Multiline graphs with separate scaling are difficult to read and should be avoided. If possible, plot all data from each test on one graph dedicated to that test.
- **Focus on facilities.** If there are several sterile manufacturing facilities, each should be reported separately. It is often informative to be able to compare different filling rooms. It may be possible to include all the graphs from the several separate tests applying to one sterile manufacturing facility on one or two pages.
- **Include all the available data on the graphs.** Most environmental microbiological data from aseptic manufacturing facilities tend towards a value of 0 cfu with occasional much higher counts that result in action limit infringements. The data presentation should allow the identification of any other intermediate condition if it should arise.
- **Include the limits on the graphs.** It is essential to know if the general pattern of data is well within its limits or if it is close to the limits.
- **Identify the locations, dates, identities and other pertinent details of action limit infringements.** This may help a "second pair of eyes" to identify a repeating problem that requires attention.

The frequency histogram is a good method of achieving these purposes. The data values (say 0, 1, 2 , 3 ... n cfu per m^3) should be plotted on the x-axis, with the number of datum points for each value plotted on the y-axis. Alternatively the percentage of the total data collected may be plotted on the y-axis. Data values may be plotted as group (say, 0 cfu, 1 to 5 cfu, 6 to 10 cfu, etc.) to improve the clarity of presentation, and the x-axis may be split for the same purpose. Examples from settle plates in an aseptic filling room and its supporting change room are given in Figure 2.4.

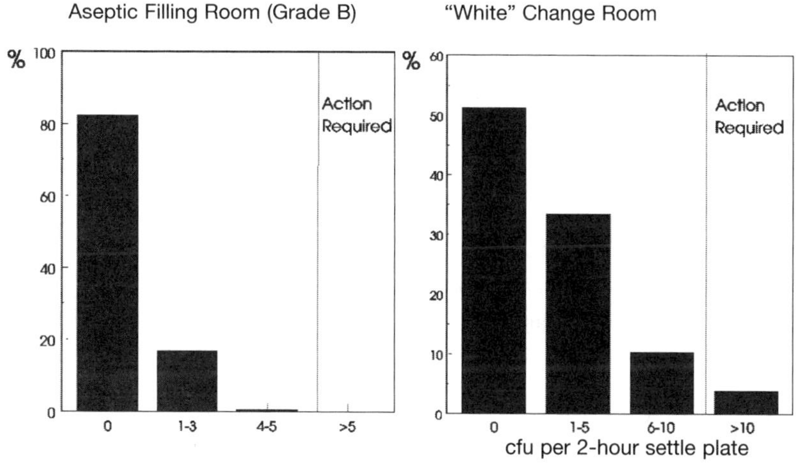

Figure 2.4. Frequency histograms of microbiological environmental monitoring data.

Data from personnel monitoring ought best to be overviewed for each individual separately because data are more likely to relate to individual practices than to general problems within the facility. In the interests of sensitivity and privacy these data should be published in overview reports with wide distribution. Any anomalous conditions or differences among operators (irrespective of whether these are within limits) should be discussed personally with area supervision, and the overviews kept on file subject to any national or local legislation relating to the individual rights of privacy.

7 DOCUMENTING ENVIRONMENTAL MONITORING

It is not easy to document environmental monitoring. There are too many themes, not merely microbiological, but also immediate physical measures. They interact and overlap, and finally integrate in an "holistic" manner. With changing standards and with local responses to the needs of individual regulators, environmental programs tend to grow erratically until documentation becomes difficult to sustain.

This chapter recommends a particular approach that many companies have found successful. This program splits environmental monitoring into "bite-size chunks" under an "umbrella" site policy. "Beneath" the site policy there is a site environmental program. Alternatively, there may be departmental environmental programs according to practical needs.

Further, there are methods unique to the environmental program (e.g., active air sampling, swabbing) and, further still, "enabling" methods and techniques that allow the program to be run properly (growth support tests, identification, etc.).

This type of approach to documentation allows personnel to be trained in specific techniques according to organisational needs, and in monitoring those parts of the facility to which they are assigned (in large facilities this may not be everywhere). It also allows for easy revision.

7.1 Site Environmental Monitoring Policy Document

This document should contain the following elements of the environmental monitoring program and concentrate on the needs of regulatory compliance.

- **Responsibilities**
  In areas of considerable complexity it is necessary to demarcate who does what, and how and when and why, or be answerable for the consequences.
  1. First there is the responsibility for preparation of the environmental policy, for ensuring that it is compliant with regulatory standards, and that local procedures are implemented in compliance with the policy. This is clearly a quality responsibility.
  2. There are then the responsibilities for "doing" the environmental monitoring. This need not be detailed at this stage, but whereas most of this responsibility would normally be held by a quality group (such as QA microbiology), any components of the environmental program that are to be done by production (e.g., personnel monitoring may be done under the supervision of a production supervisor) or by engineering (e.g., monitoring of pressure differentials or total airborne particulate) should be clearly identified here. If responsibility for aspects of monitoring lies outside the quality group, there must be some means of periodic check monitoring or audit to ensure it is done properly.
  3. There is the responsibility for training personnel in environmental monitoring.
  4. There is the responsibility for defining how data are to be recorded and reported.
  5. There is the responsibility for reviewing data and ensuring that there are appropriate reactions to all out-of-specification and atypical results. The responsibility for investigating such events, correcting them and ensuring that, they do not reoccur, may require a broader base.
  6. Finally, there is the responsibility for periodic analysis and reporting of

adverse trends in environmental data with a view to ensuring that potential out-of-specification conditions are "headed off at the pass."

- **Scope of the Environmental Monitoring Program**
  The "environmental monitoring program shall include the following measures" policy section should list those methods and techniques that yield environmental monitoring data. The units in which the data are to be recorded should be defined to allow comparison with limits set for each measure. If there is an "early warning" alert limits policy, this should be stated in this section, with mention of how such limits are to be established. For microbiological testing the policy document should include the media and incubation conditions to determine what will be recovered. For instance, if anaerobic conditions are not included, then anaerobes will not be detected.
- **Area Grading**
  The "where's" and the "when's" of environmental sampling are functions of the activities in an area according to their perceived risk to asepsis and to the degree of protection required. The *Guide to Good Manufacturing Practice for Medicinal Products* (MCA, 2002) requires that areas are graded A to D on this basis. This grading system is only mandatory in the E.U. There is no reason why sites that do not have to comply with E.U. regulations should not choose another classification system. For instance the 1987 FDA Guideline (FDA, 1987) and its proposed revisions recommends grading of areas as "critical" and "controlled." Either way, the principle is that areas must be graded with respect to the risk to asepsis. The approach to grading should be contained in the policy document.

7.2 Site Environmental Monitoring Program Document

The policy document should be very stable. If found otherwise in practice, it merits serious review. Conversely, the site program document is intended to reflect actual practice, and may be subject to more frequent revision.

- First of all the program document should contain a floor plan of the aseptic facility with the areas within the area clearly identified by their grades.
- On this floor plan the locations for environmental monitoring should be marked. The locations should be identified versus the monitoring technique to be used at each particular location (say, *T01–n* for total particle counts, *V01–n* for active microbial air samples, etc.).
- It is then quite easy to tabulate the locations against frequency of testing (e.g., batch- or time-related), and against the alert and action limits to be applied.
- The program document should identify how media and equipment should be taken into aseptic areas, how samples should be labelled, and how they should be accounted for and reconciled for incubation, read out and review.

- The program document should ideally contain a flow diagram of responsibilities and interfaces.
- The program document should contain copies of all formal report forms.
- The program should contain directions about what to do when limits are exceeded.
- The program should state which of the microorganisms recovered in the program should be identified. The extent of identification may be quite different from one area grade to another.

The environmental monitoring program document should describe everything the personnel involved in undertaking environmental monitoring need to know. There are also important things that it should not contain, best referenced to other formal documents that can be self-contained. These other documents should, in common with the site environmental policy document, be quite stable. In "process" order these document should separately address the following "enabling" processes:

- Media preparation
- Growth support tests
- Total particle counters and how to use, maintain and calibrate them
- Microbiological sampling methods (e.g., active air sampling, swabbing, settle plates, etc.). These can be addressed in one document or split out into one document for each technique
- Identification of microorganisms
- Handling of out-of-specification and atypical results
- Conduct and reporting of trend analyses

This list is not definitive.

The author favors the principle of a documentation system that allows environmental monitoring to be addressed in a practical way, with procedures compartmentalized and focused for the likely tasks done by different personnel.

Thus documents will be facilitated as training aids and as day-to-day guidance, in the ongoing quest for safe pharmaceutical manufacturing environments and patient benefit.

REFERENCES

Agalloco J. Qualification and validation of environmental control systems. *PDA Journal of Pharmaceutical Science and Technology*, 50: 280–289, 1996.

Akers JE. Environmental monitoring and control: proposed standards, current practices, and future directions. *PDA Journal of Pharmaceutical Science and Technology*, 51: 36–47, 1997.

Benbough JE. *The Sampling Efficiency of the Biotest RCS PLUS Air Sampler.*

Biosafety Test Section, Division of Biologics, PHLS Centre for Applied Microbiology, Wiltshire, U.K., 1992.

Food and Drug Administration of the United States Department of Health and Human Services (FDA). *Guideline on Sterile Drug Products Produced by Aseptic Processing*. Rockville, MD: Center for Drugs and Biologics, 1987.

Kaye S. Efficiency of "Biotest RCS" as a sampler of airborne bacteria. *Journal of Parenteral Science and Technology*, 42: 147–152, 1988.

Ljungqvist B. and Reinmuller B. Some aspects on the use of the Biotest RCS air sampler in unidirectional air flow testing. *Journal of Parenteral Science and Technology*, 45: 177–180, 1991.

Medicines Control Agency. *Rules and Guidance for Pharmaceutical Manufacturers and Distributors*. London: The Stationery Office, 2002.

Parenteral Drug Association (PDA). *Validation of Aseptic Filling for Solution Drug Products, Technical Monograph No 2*. Bethesda, MD: Parenteral Drug Association Inc., 1981.

Parenteral Society. *Technical Monograph No. 2 — Environmental Contamination Control Practice*. Swindon, U.K.: Parenteral Society, 1990.

PDA. Technical Report No 24. Current practices in the validation of aseptic processing. *PDA Journal of Pharmaceutical Science and Technology* 51 Supplement S2, 1997.

Parks SR, Bennett AM, Speight SE, Benbough JE. An assessment of the Sartorius MD-8 microbiological air sampler. *Journal of Applied Bacteriology*, 80: 529–534, 1996.

Pendlebury DE, Pickard D. Examining ways to capture airborne microorganisms. *Cleanrooms International*, 1: 15–30, 1997.

Sykes G. The control of airborne contamination in sterile areas. In *Aerobiology — Proceedings of the 3rd International Symposium*, ed. Silver I.H. London: Academic Press, 1970.

Whyte W. Sterility assurance and models for assessing airborne bacterial contamination. *Journal of Parenteral Science and Technology*, 40: 188–197, 1986.

Whyte W, Niven L. Airborne bacteria sampling: the effect of dehydration and sampling time. *Journal of Parenteral Science and Technology*, 40: 182–188, 1987.

Chapter 3

Media Fills and Their Applications

Nigel Halls

CONTENTS

1 INTRODUCTION

Along with many other publications in good manufacturing practice (GMP) our purpose is to guide readers to a starting point, from which they can then progress to

0-849-32300-2/04/$0.00+$1.50

a deeper understanding of the nature of microbiological contamination in drug production.

The particular purpose of sterile drug manufacture by aseptic processing is the avoidance of microbiological contamination. Proper clean-room design, engineering and operation make the probability of finding nonsterile units in populations of sterile dosage forms often immeasurably low.

Some contamination of aseptic filling lines and their surroundings is, however, usually unavoidable. The probability of finding microorganisms contaminating aseptic filling rooms is higher than that of finding contaminated dosage forms. The probability of finding microorganisms contaminating support areas and change rooms is even higher.

The purpose of media fills and related operations, such as environmental monitoring, is to obtain an index of the typical levels of microbiological contamination occurring in aseptic manufacturing and their support areas. These indices of typicality are used as comparators to identify unusual events, which may indicate occasional or persistent lapses in contamination control. The levels of contamination tolerance vary from situation to situation.

Here we help unravel any mystique surrounding microbiological contamination, for readers at all levels with an interest in the pharmaceutical industry.

2 MEDIA FILLS: PURPOSES

Media fills, broth fills, simulation trials and so on, are all synonymous names for an exercise undertaken as part of the validation of a new aseptic process, and as a frequent validation review thereafter. Regulatory pressure to repeat media fills at six- and even three-month intervals, is gradually merging media fills into routine environmental monitoring programs.

The purpose of the media fill is to provide a measure of the likelihood of microbiological contamination arising in particular aseptic processes. A placebo is substituted for the product, and is processed in an identical manner identical to the processed product.

In its simplest form, an aqueous liquid microbiological growth medium is substituted for an aqueous liquid product. The medium is incubated, the number of contaminated versus uncontaminated units are scored, and decisions made based on the number or proportion of contaminated units, and from the identities of the contaminating microorganisms.

It should initially be emphasized that media fill results do not provide an index of the *probability* of nonsterile product units in product populations. In other words, they do not represent a measure of the Sterility Assurance Level (SAL) achieved for any particular aseptically filled product. This conceptual difference between the proportion contaminated in a media fill and the SAL of sterile products is generally poorly understood.

In properly conducted media fills the aseptic process is simulated exactly as it would be carried out routinely. The only difference is the use of a placebo to replace a pharmaceutical product. An aqueous placebo is used to simulate aqueous liquid dosage forms, a solid placebo to simulate sterile solid dosage forms, and something with similar rheological characteristics to an ointment to simulate ointments.

Table 3.1 compares the composition of the placebo most commonly used for aqueous liquid media fills, Tryptone Soy Broth (TSB), with the formulation of a typical aqueous injection. There is no coincidence. TSB is widely used as a placebo for media fills as a general-purpose microbiological growth medium, in which a broad spectrum of types of microorganisms is expected to survive and multiply.

Table 3.1 Comparison of TSB with an Aqueous Injection

	TSB (g/l)	Aqueous Injection (g/l)
Drug substance		28
Casein	17	
Soy bean meal	3	
Dextrose	2.5	
Phenol		5
NaCl	5	
K_2HPO_4	2.5	
KH_2PO_4		1
Na_2HPO_4		2.4

The injection described in Table 3.1 has been formulated for completely different purposes, most significantly containing a preservative (0.5% phenol) for the express purpose of inhibiting the survival and growth of microorganisms. The proportion of contaminated units found in media fills is based on the process in which this product is filled. This is arguably the worst SAL for this aqueous injection, or any other aqueous injection filled in the same process. Frankly, however, this is unlikely to bear a major resemblance to the real probability of finding a nonsterile unit in a manufactured population, batch or lot.

Another example to emphasize the same point — that media fills simulate process contamination and not SALs — is evident when different dosage forms with different formulations, drug substances, preserved and nonpreserved, etc. are manufactured in the same filling process. It is not feasible that they should all produce the same SALs, because of the effects of their formulations on contaminant survival. But it is only usual to perform one set of media fills for each process and to obtain only one index of the probability of contamination in the process.

3 MEDIA FILLS: PLACEBOS

The most commonly used placebo for media fills is TSB used to simulate aqueous injections. It is a reasonably good all-round, general-purpose microbiological medium, which can support growth of aerobic bacteria when incubated at temperatures in the range 20–35°C. Equally, it is a reasonably good medium for supporting the growth of yeasts and fungi, when incubated at 20–25°C. It is the recommended test for sterility in all of the major pharmacopoeias.

However, many microorganisms will not readily grow in TSB and some will not grow at all. It is a good recovery medium for Gram-positive and human commensal-type bacteria, but not the best recovery medium for Gram-negative bacteria. The latter grow better with lower nutrient concentrations, and at lower incubation temperatures, than those recommended in the pharmacopoeias for the test for sterility.

TSB is not the best recovery media for yeasts and fungi. A mycology specialist would not use TSB as the first choice for surveying an environment for yeasts and fungi. It the not best recovery medium for anaerobic and microaerophilic microorganisms such as the common skin commensal, *Propionibacterium acnes*.

Why is TSB used so widely if it displays so many limitations? The answer is, quite simply, that it is a compromise medium, commercially available, uncomplicated and robust. It is supported by the reflected authority of the pharmacopoeias. Most importantly, it has become the industry standard.

In the interests of academic science, it could be desirable to use a better medium for media fills, or more types of media for each media fill, but in practical terms there is little benefit. The media fill is not an exhaustive search for every microorganism that could be contaminating an aseptic process — it is a "snapshot" in time with a recognised and limited "focal range."

A wider variety of placebos is used for solid dosage forms. Generally, the placebo is filled into the unit containers and then TSB is added, either on- or off-line. It is possible to add the TSB before the placebo, but it is not general practice. The placebo is dissolved in the TSB and incubated.

The chosen placebo should have similar flow characteristics to the product or products that it has been chosen to represent. If it does not have these similar characteristics, it might be effectively impossible to simulate the intended process.

It must be sterilizable. Gamma irradiation is the method of choice for sterilizing solids provided they have a low moisture content; it is unlikely to induce chemical or physical changes through radiation. Irradiation is reliable and penetrative through bulk quantities.

The placebo must be soluble in TSB, and must not inhibit the growth of microorganisms. The practical application of this principle is complicated by the fact that the amount of space available for TSB in each container is restricted by the amount of placebo already added. This in turn must be sufficient to stimulate the process.

Polyethylene glycol, mannitol and lactose are the most widely used placebos for solid dosage forms.

Performing media fills on ophthalmic ointments is a nightmare. Placebos are based on TSB made viscous by the addition of a substance such as carboxymethyl cellulose at about 65 g/l, although this concentration may differ according to the process settings applicable to the range of ointments being simulated.

The nightmarish aspects of ointment media fills are three-fold.

First, there is cleaning up behind them. Actual ointments generally present a sticky mess, which is difficult but obviously not impossible to clean from production equipment. *But*, add a microbiological growth medium to that sticky mess, and cleaning becomes highly critical, especially if the creation of foci for microbiological growth in the equipment and in the facility is to be avoided — as it must. Good clean-room practices are difficult to maintain in ointment media fills.

Second, ointment tubes are rarely transparent, therefore inspection of thousands of placebo-filled tubes for growth after incubation is difficult. The tubes are opened and squeezed out, although some users of plastic tubes special-order transparent tubes purchased solely for media fills.

Third, microorganisms grow as colonies in carboxymethyl cellulose-thickened TSB rather than displaying a general opacity. Carboxymethyl cellulose-thickened TSB is not a clear transparent medium in which colonies can be easily discerned. This is addressed by inclusion of a metabolic indicator, such as 2,3-tri-phenyltetrazolium chloride in the medium, at or around 0.0025%. Tetrazolium chloride is a metabolic indicator that changes to a red or purple color when microorganisms respire.

4 MEDIA FILLS: SIMULATION OF PROCESSES

Regulatory literature abounds with restrictions that have been created with *typical* aseptic processes in mind — the pharmaceutical manufacturing industry is populated by responsible citizens baffled by these rules, and how they should be applied to their *atypical* processes.

Most, if not every, aseptic process is unique. Even in the same factory, two lines set up for the simplest process such as filling liquid products into ampoules could significantly differ.

The general principle of media fills is that the process should be simulated in a way that addresses *every* risk of microbiological contamination that could occur in practice, i.e., the process must be conducted exactly as in routine operation. In reality, usually some compromises are made specifically for media fills.

Although aqueous liquids are frequently portrayed as the typical product for generalizations on media fills as in Annex 1 of the E.U. Guide (CEC, 2002), they have their own complexities in terms of process simulation not shared by solid dosage forms or ophthalmics. In this treatment of process simulation, solid dosage forms are given as the initial example, followed by aqueous liquids, and then by consideration of more specialised applications, e.g., lyophilization, aseptic blending, and so on.

4.1 Simulation of Solid-Dosage Form Aseptic Filling Processes

Figure 3.1 very simply describes aseptic filling of a solid-dosage form into vials by two different but broadly similar technologies. In the first, the empty vials are depyrogenated in a double-door oven and loaded onto the filling machine; in the second, the empty vials are depyrogenated in a tunnel linked to the filling machine. Other than that, the processes are the same: rubber closures are sterilized in double-door autoclaves, the bulk sterile dosage form is brought into the filling room via an air-locked hatch, and personnel are required to enter the filling room to service and operate the processes.

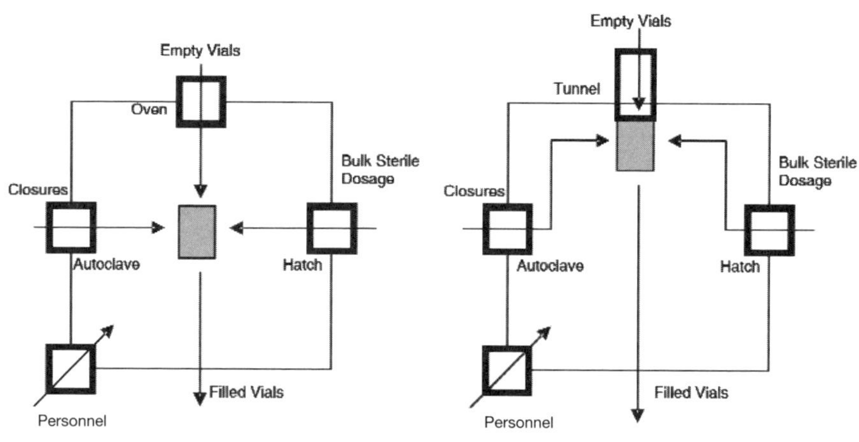

Figure 3.1. Simplified representation of aseptic filling of a solid dosage form into a vial.

For media fills, the placebo is substituted for the bulk sterile dosage form in exactly the same type of container. It is brought in through the hatch and taken and connected to the filling room and filled. TSB is then added to each vial. This may be done using an on-line liquid filler, which adds an extra aseptic stage to the filling process, or off-line. If the medium is added off-line, the time between filling the placebo and adding the medium becomes critical.

The filling process is then run as identically as possible in routine practice, with the following exceptions.

1. Any inert gas (e.g., CO_2 or nitrogen) used to fill or sparge the vial headspace should be disconnected, or compressed air should be substituted for the gas. The principle of the use of placebos and culture media is to create conditions where there is the greatest possible likelihood of recovering any contaminants present. Most contaminants likely to be present in pharmaceutical manufacturing environments metabolise aerobically and the creation of anaerobic conditions in the headspace above the media would decrease the probability of recovering these aerobes.
2. The weight of placebo added to each vial need not necessarily be the same as the weight of the product. Typically they are identical for small fills. With larger fills it is not always usual practice to replicate the exact weight of the product, as long as the filling speed is adjusted to leave the vials open under the filling heads for the same time as they would be in routine filling. The principal reasons for doing this are in connection with media.
    - The concentration of placebo in media must not be so high as to inhibit microbial growth. The smaller the weight of placebo present per vial, the easier this is to achieve. Polyethylene glycol is not inhibitory to microbial growth in TSB in concentrations of up to 100 g/l.
    - Sterilization of microbiological media for media fills is a logistics problem faced by many microbiological QA laboratories. The greater the amounts of media required, the greater the problem. This can be minimized by using smaller amounts of media with smaller weights of placebo.
3. *All* contaminating events permitted in a specific process must be simulated in the time that the media fill is running, even though some may be infrequent events. Before media fills are run to validate a new process, and perhaps where there is little past experience of filling solid dosage forms, the operational Standard Operating Procedure (SOP) should be carefully scrutinized and the process "brainstormed" to prepare a list of potential contaminating events. This can be checked off during the media fill at the time they are simulated.

    With existing processes, where personnel or wear and tear may have introduced informal changes to the process, it is sensible to repeat the "brainstorm" with the operational personnel periodically. Observe the operational process closely over several shifts, noting what happens and how often. Typical contaminating events include, but are not restricted to:

    - Setup of the filling equipment prior to commencement of fill.
    - Placebo-container changes. This is usually a manual process and each time it happens there is some risk of operator contamination.
    - Replenishment of closures in the closure-hopper. This is also generally a manual process.
    - Replenishment of vials in the vial-feed if this is a manual operation from a depyrogenating oven; this is not an issue with tunnel depyrogenation.

- Filling-machine adjustment at the beginning of the process and any adjustment that might be necessary in response to, for example, weight checks. These in-process machine adjustments must be simulated, even though they may not be necessary in the actual media fill.
- Filling machine stoppages.
- Removal of vials that have fallen over, etc.
- Off-loading of stoppers from autoclaves.
- Personnel shift changes and other occasions where personnel may leave or enter the filling room.
- Microbiological monitoring.

4. The most potent source of contamination in aseptic processes is personnel. It is important that any potentially contaminating event associated with manual intervention is addressed through each of the human variables. Each aseptic operator should be required to actually perform or simulate the performance of each potentially contaminating event in each media fill. In order to do this reasonably, it is customary to split human intervention potentially contaminating events into categories: "minor," "major and standard," or "critical, intermediate and standard." The choice of name for the categories is discretionary, but it is often regarded as unwise to speak to regulatory agencies using the term "minor" in relation to an aseptic intervention. It should be ensured that each aseptic operator performs all of those within the most serious category for each media fill. Less serious interventions need only be addressed by the "team," as distinct from each member of the team.
5. The media fill need not run over a complete shift, just long enough to fill a statistically significant minimum number of units. It needs to be enough to be able to simulate all of the potentially contaminating events, and to address the potential for contamination to build up over time.

   The contents of each vial are only likely to be contaminated while the vial is open and its contents unprotected; this will be for a matter of seconds only in most aseptic processes. Irrespective of shift length, each vial is still only open for a few seconds. Admittedly there is a possibility of the concentration of contaminants increasing in a clean room over the time it is manned and operational, but this is addressed in routine liquid media fills at the end of a normal production run, with the personnel who have been working in the area. The only exception to this practice is for antibiotic filling, where it is important that all antibiotic traces are cleaned out of the filling equipment and the filling room before the placebo is filled. This is to prevent the antibiotics from inhibiting recovery of microorganisms in the medium. It is advisable to use personnel who have completed or are near the end of a shift on another filling line to simulate antibiotic filling, to simulate any "sloppiness" in aseptic technique that may arise from tiredness.

   A more rigorous approach may be demanded to the validation of the time a sterile "setup" may be left on a filling machine, especially if filling is done on a campaign basis over more than one day. There are several possible approaches to this.

- Several thousand units may be filled with placebo and medium after start-up. Unless the filling machine is sterilize-in-place (SIP)-equipped to point-of-fill, machine setup and aseptic assembly of presterilized product contact parts is surely one of the times of greatest contamination risk. Thereafter the machine may be "held sterile" for a period of hours or even days, and then several thousand more vials filled with media, with all interventions included or simulated. Thus the three major risks — setup, interventions and time-related factors — are all taken into account.
- Alternatively, several thousand units may be filled with placebo and medium after start-up, and then the machine may be "run dry", i.e., with no addition of placebo or TSB for as long as necessary, with operators freeing jams and simulating sample removal, as usual. The vials may then be filled with placebo and medium as before.
- The third alternative is for the machine to run placebo for the whole of the campaign length that is to be validated. Medium is, however, only filled for the first and last several thousand and after any serious interventions during the "placebo-only" period.

4.2 Simulation of Aqueous-Liquid Aseptic Filling Processes

Figure 3.2 describes aseptic filling of an aqueous liquid into ampoules, in the same simplistic way as in Figure 3.1.

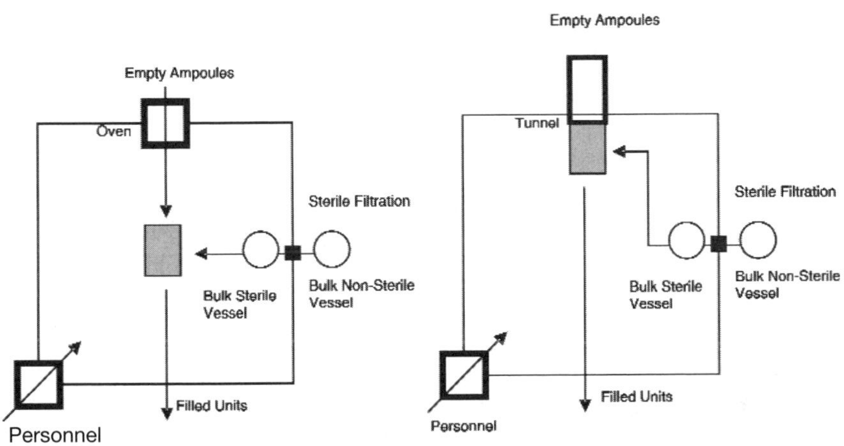

Figure 3.2. Simplified representation of aseptic filling of a liquid dosage form in ampoules.

At first glance it seems that aseptic filling of liquids is a less complicated example than solid dosage forms, as there is no need for an extra filling stage — TSB is both placebo and recovery medium. Indeed, all of the provisions to the conduct of media fills also apply.

In summary:

- Inert gas sparging should be replaced with compressed air.
- The volume of medium filled may be reduced. In a 1996 Parenteral Drug Association (PDA) survey of aseptic manufacturers, some 34% respondents did not fill the same volume of media as they filled of product in routine production (PDA, 1996).
- All contaminating events must be simulated. In modern, high-speed, tunnel-linked ampoule filling lines, this often results in as many as 10,000 or 20,000 units filled just to give enough time to simulate everything.
- The duration of the media fill has to be long enough to simulate everything needed but not so long as to create problems with incubator space.

A divide has arisen among sterile liquid manufacturers. Traditionally, aqueous-liquid media fills were done by taking vessels of autoclave presterilized TSB into the filling room, connecting them one by one to the filling line, and then filling the ampoules or vials, etc. Some manufacturers, however, interpret regulatory pressure to "simulate the whole process" to mean that they must take dehydrated culture medium as their starting point, make it up in their manufacturing areas, pass it through the process sterilising filters and then connect to the filling room and fill ampoules or vials. Both approaches have some advantages and disadvantages.

The origins of the "traditional approach" lie in older, slow-speed technology, when a regulatory-satisfactory media fill could often be achieved by filling as few as 1000 units. There would be sufficient laboratory autoclave capacity to sterilize sufficient media in aspirators, or large vessels that could then be brought to the filling machine.

One aseptic connection would have to be made between the media vessel and the filler. However, in routine operation, there would most likely be other additional aseptic connections, e.g., between the downstream side of the sterilizing filter and the sterile holding vessel. Very few older processes have the SIP systems addressing the whole line — from filters to filling needles — now developed for newer processes. Inevitably, aseptic connections would also be required and probably not be simulated by the traditional approach to the media fill. Conscientious manufacturers might simulate these aseptic connections separately to the media fill; others would ignore them.

With the advent of high-speed filling lines and the need for larger numbers of filled units, laboratory autoclave capacity often became a limiting factor in complying with regulatory requirements on numbers of units filled. The question was inevitably asked as to why autoclave sterilization was necessary for media, when there was a perfectly good sterilizing process (filtration) used for the product. So filtration through the production filtration setup came a fairly commonplace practice.

1. **By taking dehydrated media through all the stages of dispensing and compounding in production vessels, every potential for contamination of the media is taken into account.** There is some confusing logic in this contention.
   - Dehydrated microbiological media is most usually heavily contaminated with microorganisms reaching levels of around 10^4 colony-forming units (cfu)/g. Raw materials for aseptic manufacture are invariably specified to be within standards of contamination of no more than 10^3 cfu/g and rarely ever approach those limits. Compounding areas must be restricted and microbiologically controlled — they are a medium-level clean room. Operators in compounding areas must wear dedicated footwear, clean overalls, head covers and gloves. At least twice a year, in the name of QA and regulatory requirements, the notion of bringing nonsterile, dehydrated microbiological media through these areas makes a mockery of the other enforced controls. If simulation of the filtration process is thought to be valuable to the media fill, it is sensible to have the dehydrated media sterilized by gamma radiation, or for prepared media to have been autoclaved before it is brought into the compounding areas.
   - The media fill is intended to detect weaknesses in aseptic processing. Compounding is intended to be sufficiently clean to prevent increases in contaminants or of their byproducts (e.g., endotoxins) resulting from conditions in the manufacturer's premises, but it is not an aseptic process. The media fill should not be seen as an instrument for detection of problems in nonaseptic manufacture; there are simpler and more straightforward methods to achieve that end.
2. **The media follows exactly the same route as the product and is therefore an exact simulation of the process, including the risks associated with sterile filtration.** Indeed there is some contention that the media fill validates sterile filtration — it does not.

There is a totally independent regulatory requirement for sterile filtration to be validated by a bacterial challenge test that is specified in detail and relates to the way filters, particular microorganisms in particular concentrations, and specific products interact. Sterilizing filters are not intended to retain microorganisms at particularly high challenge levels and at the viscosity of microbiological media. The newer approach to simulating the challenges to the product is probably fairer than the traditional approach, but its limitations must be recognized.

4.3 Simulation of Processes Involving Aseptic Bulk Compounding Before Filling

Some sterile products need to be compounded aseptically, e.g., suspensions. Some antibiotic solid-dosage forms require blending with a carrier. Each particular case is likely to be different. There may be compounding of two liquid phases, both of which have been passed through bacteria-retentive filters; compounding of two solid phases both of which enter the filling room through pass-through hatches; or compounding of liquid and solid phases.

The general rule for media fills is that the aseptic compounding needs to be included in the simulation.

4.4 Simulation of Lyophilization Processes

Those sterile dosage forms that are stable only for a short time in solution are frequently marketed in lyophilized presentations (see Figure 3.3).

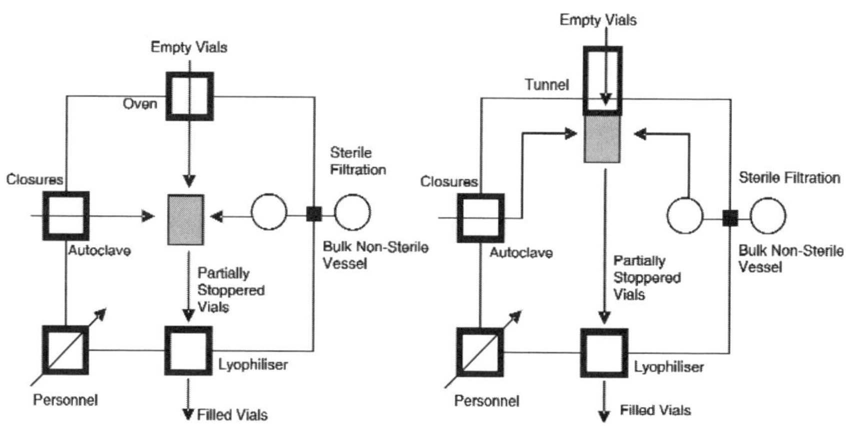

Figure 3.3. Simplified representation of aseptic filling and lyophilization.

The process is more complicated than standard vial filling, although it may involve many items of common equipment. Vials are aseptically filled in the normal way, but the closures (which are of a special design) are not fully inserted.

The filled, partially stoppered vials are "trayed," taken and loaded into a lyophilizer. The traying and transfer of the vials from the filling machine to the lyophilizer may be done manually or automatically (e.g., by robotics and automatic goods vehicles). Irrespective of the means, the contents of the vials are vulnerable to contamination while they are only partially stoppered.

Within the lyophilizer the liquid in the vial is frozen and a vacuum drawn. The water from the solid (frozen) phase sublimes directly to vapour, and the dosage form dehydrates. At the end of the cycle the vacuum is broken and the closures are automatically rammed home. The main vulnerability of the process to microbiological contamination is clearly at the point where the vacuum is broken and air enters the lyophilizer and the vials. Replacement air must be filtered sterile, but other undiscovered means of air contamination from leaks, bypasses, etc. cannot be discounted.

What should and what should not be simulated?

1. The aseptic filling process should be simulated exactly as any other vial-filling process. However, since the closures and the vials may differ, attention should be given to simulating any activities that are peculiar to filling lyophilized vials as distinct from liquid-filled vials. There may be a greater frequency of intrusion to free blocked closure chutes, or to remove vials that have fallen over. Any such difference will be unique to the particular process and have to be determined empirically.
2. The traying and transfer process should be simulated exactly.
3. The lyophilization process itself must not be simulated exactly.
    - The freezing of vials and the formation of ice crystals is inimical to microorganisms. Those who argue that lyophilization is one of the most frequently used methods of preserving microorganisms and is therefore not inimical, have clearly never experienced the difficulties that microbiologists endure and overcome to ensure viability when using lyophilization for preservation purposes. Freezing should not be simulated: 24 of 26 manufacturers using lyophilization who responded to the PDA's 1996 survey of aseptic manufacture claimed not to freeze their media fill vials (PDA, 1996). If there is danger of unfrozen media foaming over under vacuum and thus contaminating the lyophilizer, it may be necessary to double the size of the media fill. Simulate all of the risks up to and including loading of the freeze dryer in one-half of the media fill that is not frozen, and then simulate the subsequent risks with the rest of the filled vials that are passed through the complete process including freezing.
    - A complete vacuum as specified for the lyophilization process should not be drawn. In addition to the technical difficulties of foaming, which would happen if a complete vacuum were to be drawn over the liquid rather than solid-phase dosage form, consideration should be given to any fluid loss from the media and its effect on the viability of microorganisms

> and the ability of the media to support microbial growth. These are two separate issues. After some concentration the media may still be able to support the growth of microorganisms, but injured microorganisms may have died as concentration took place. Typically, a partial vacuum of say 20 to 28 inches Hg is drawn, held for about two hours and "broken." Conscientious simulators of worst-case conditions may repeat this process although it is not typical of routine practices.

Some companies perform complete simulation of the lyophilization process from filling, through transfer, to lyophilization. Others may split the process into three simulations to help provide a clearer focus on what might have gone wrong if contaminated units result from the media fill. The decision as to which approach to take or how to develop a responsible combination of the two approaches is a matter of judgement. A balance has to be struck between regulatory pressure to simulate the process as closely as possible, and the need (also pursued rigorously by regulatory inspectors) to diagnose the source of contamination accurately enough to implement satisfactory corrective or preventive actions.

5 MEDIA FILLS: MICROBIOLOGICAL CONSIDERATIONS AND CONTROLS

The "ownership" of media fills should properly lie with the management of the aseptic process. Properly, media fills should be scheduled into the manufacturing program in the same way as a routine filling activity except that the product is units filled with media. In practice, this ownership tends to be held jointly between production and microbiological QA. Numerous microbiological considerations and controls must be complied with for a media fill to be fit for its intended purpose within the QA program.

5.1 Growth Support and Sterility Controls

The first responsibility in any microbiological exercise that is expected to produce "no growth" results, and for which no growth is the favorable condition, is to ensure that the medium is capable of supporting growth. Maintenance of aseptic clean rooms must ensure that only materials that can be safely presumed to be sterile should be permitted entry.

Growth supportiveness of the media should be verified before use. It should also be checked after it has been in contact with the filling equipment and the product containers. This is to ensure that traces of product, antibiotic, detergent, disinfectant, etc. in antimicrobial concentrations have not been passed into the media from any one of these or other production-related sources.

Sterility of media is best verified by preincubation. This is best done outside of aseptic areas. The prospect of having, for example, 50 litres of microbiological media becoming heavily contaminated through each hour of preincubation within an aseptic filling room should be avoided.

When prepared media are autoclaved for media fills a sample is usually aseptically withdrawn for growth support checks. These can then be simultaneously conducted with preincubation of the media in a laboratory incubator to verify its sterility. Both results are obtainable before the media need be taken into the clean room.

If the only sterilization of media is on-line filtration, an aseptic sample may be taken from the sterile holding vessel for growth support checks. The media should properly be held until these results are obtained, but the risk of contaminating the filling room by preincubation therein is something some companies prefer to avoid. In such a case a risk may have to be taken to fill media for which there is no prior supportive evidence of either sterility or of its ability to support growth. There may be additional risks of contaminating the medium by moving it out of the aseptic filling room for preincubation and back in again for filling. These may have to be tolerated.

A second growth support check should be done on filled units. In principle these may be taken and tested at the beginning or end of the incubation of the media fill. In terms of managing and scheduling it would be best to take them at the beginning. This eliminates taking the whole of the media fill incubation period plus some days before ascertaining that the media was satisfactory. However, in response to a well-known but informal regulatory view that this practice may result in taking the very units that might be contaminated out of the trial, growth support on filled vials is most usually done at the end of the incubation period of the complete set of filled units.

The medium that is universally used for media fills is TSB, because it is used in the pharmacopoeial test for sterility. This is the usual point of reference for the microorganisms and the conditions that should be applied to growth support checks.

Table 3.2 shows the current *United States Pharmacopeia* (*USP*) and *European Pharmacopoeia* (*PhEur*) requirements for TSB medium growth support when used for the test for sterility; however, the control cultures applying to TSB are cited only for the 20–25°C incubation condition. Media fills may be incubated at two temperatures, 20–25°C and 30–35°C. It is therefore good sense to replicate the growth support test across the two temperature ranges. All pharmacopoeially recommended microorganisms listed in Table 3.2 should grow profusely in both temperature ranges with seven days' incubation from an initial inoculum of 10 to 100 cfu. Separate media samples should be inoculated with each culture.

In addition to the pharmacopoeial media growth support control cultures, many regulatory agencies insist on at least one isolate from the manufacturing environment being used as a media control.

The logic is that if the TSB is intended to recover microorganisms inhabiting the manufacturing environment, it should be shown to have the ability to support the

Table 3.2 Microorganisms Required for Sterility Test (Media Growth Support Checks in *USP XXVI* [2003] and *PhEur 4th edition* [2002] (ATCC Numbers Only are Shown for Convenience)

Medium	*USP XXVI* (2003)	*PhEur 4th edition* (2002)
TSB at 20–25°C	*Bacillus subtilis* (ATCC 6633)	*Bacillus subtilis* (ATCC 6633)
	Candida albicans (ATCC 10231)	*Candida albicans* (ATCC 2091)
		Staphylococcus aureus (ATCC 6538P)

growth of those environmental microorganisms. Local microorganisms could be frail, injured, disinfectant-damaged, etc. and therefore could be more difficult to recover in TSB than the pampered, well-nourished subcultures from the culture collection.

Conversely, the local environmental isolates used for media controls will have most likely been maintained in a local culture collection for several months at least, and will probably have recovered from any physiological damage associated with stressful local conditions. It may seem cynical, but it is probably true to state that few microbiologists would choose a local environmental isolate for media fill control because it is difficult to grow or is slow growing in laboratory culture.

Irrespective of these doubts and compromises, local environmental isolates are recommended for media control. The chosen isolate should be changed periodically so that it can be related to the current rather than the historical microflora of the manufacturing environment.

Where antibiotic filling processes are simulated, ensure that at least one of the growth support control cultures is sensitive to the antibiotic, to provide the most sensitive information on the success of the clean-up process.

The preparation of control cultures should be clearly specified in laboratory documentation, and records of subculturing maintained. The FDA prefers that working control cultures be separated by no more than five generations from their national or international culture collection origins. This limits the potential for mutation.

Low inocula must be used in media control, because the intention is to recover microorganisms when they are present only in low numbers. The pharmacopoeias interpret low numbers to mean between 10 and 100 cfu per inoculum. The reference condition for this is surface culture on Tryptone Soy Agar incubated at 30–35°C for at least 48 hours.

5.2 Foul-Up Controls

Microbiological controls are not limited in their value to the qualification of the media fill. They should also be set up to expose any problems of their own creation. Microbiological QA exists to identify production problems and to assist in their resolution — it should always be wary of creating problems of its own making. Any activities associated with microbiological control of media and any laboratory manipulations that do not exist in routine manufacturing practice should be examined critically. This is best done by detailed analysis of the ways in which particular media fills are organised. Some examples are given below.

1. Aseptic sampling from bulk media is a serious vulnerability. It is all too easy for the bulk to be contaminated when growth support samples are withdrawn. It is not inconceivable that the outcome would be for the media fill to be contaminated, probably over several filled units, as a result of the contaminants distributed throughout the bulk and possibly proliferating before all of the media are filled. If possible, the bulk vessel should be incubated at the same time as the media fill. Contamination of the bulk invalidates the media fill and creates unwanted pressure attempting to diagnose production problems that are not of production's making. Regulatory agencies would always expect a media fill in which the incubated bulk was found to be contaminated to be repeated, irrespective of the quality of the results from the filled containers.
2. In some cases, solid-dosage form media fills require the addition of the recovery medium to the placebo-filled containers (usually vials) off-line in a laboratory. This requires media in bulk, and some apparatus, most often an automatic or repeating syringe, for transfer to the placebo-filled units. The vulnerabilities are for the bulk to be contaminated when the microbiologist aseptically assembles the transfer apparatus, and for the transfer apparatus to become contaminated over the period in which it is used. In this type of media fill it is easy to retain and incubate the bulk container. A sample of the first and last media passed through the transfer apparatus before any placebo-filled units are filled should be injected into a sterile vessel and incubated. Usually the placebo-filled units are interspersed with sterile sealed containers at regular intervals. The sterile sealed containers are filled with TSB in the same way as the placebo-filled containers, intended to disclose any transfer apparatus contamination as close as possible to the stage in media transfer when it happened. The frequency of interspersion of sterile containers is a matter of judgment; it may be every third, fifth or tenth unit according to the degree of confidence in the skills of the microbiologists adding the media. Irradiation is the recommended method of sterilization because sealed empty vials are quite difficult to sterilize by autoclaving. The discolouration obtained in most grades of glass as a result of exposure to gamma radiation is a convenient feature for distinguishing sterilised from placebo-filled units.

5.3 Environmental Monitoring and Media Fill Observation

Microbiological monitoring is a potential source of contamination that must be simulated in media fills, as one of the best ways of diagnosing the source of contamination arising in media fills. It should always be assumed that management is anxious to know what, where and why media fill contamination has arisen, in order to decide on appropriate corrective and preventive actions and improve their processes.

For this reason it is advisable to have intensive microbiological monitoring over the period of the media fill. The potential advantages outweigh the disadvantages. Microbiological environmental monitoring should be intensive. Where intensive microbiological monitoring may be routine practice, applied over a number of locations on a matrix basis, the practice during media fills should be for all locations to be monitored.

It is also advisable that the media fill is observed by a person who has been trained in asepsis and is familiar with the filling process. Detailed notes should be taken describing what and when is happening, particularly anything unusual. The observer may provide an independent verification that the listed contaminating events have been simulated. It is useful if the "traying" of filled units can be related to the times of filling.

Some regulatory agencies indicate a preference for media fills to be recorded on videotape. This is the best way of proving that fraudulent claims regarding the conduct of the media fill are not being made. Conversely the video camera rarely has the peripheral vision and the variability of focus of the human observer. For information purposes the video recording has to be done intelligently, but the risk is that regulatory investigators may become more interested in what the camera person may not have recorded, than what they have focused on. A fixed camera focused at point-of-fill gives no information about the risks attendant upon, for example, unloading autoclaves, replenishing stopper bowls, etc. After all, although contamination of the product unit may only happen at point-of-fill, who is to say that the contamination did not come from a stopper that was itself contaminated by an operator unloading an autoclave?

6 MEDIA FILLS: INCUBATION

Filled units must be incubated as soon as possible after filling. Regulators, the FDA in particular, are anxious that all units are incubated (with the exception of those without caps, obvious cracks, etc.). This is intended to include those "perfect" units that may in practice never be released, e.g., units cleared off the line after a stoppage or similar event. Regulators prefer incubation for information purposes, with indirect rather than direct impact on the success or failure of the trial.

When media are added in the laboratory after a solid placebo is filled, it is critical that the interval between filling the placebo and adding the medium is as short as possible. This prevents contaminating microorganisms dying off in the placebo. The argument against this, premised on genuine product contaminants dying off within the seven or 14 days' sterility test quarantine before release, is not valid. The media fill is intended to disclose process contamination, not the probability of nonsterility in product within its marketed shelf life. The maximum interval between filling and media addition should be validated by inoculation of the placebo and then tracking recoverable survivors over time.

Incubation of media fills is almost universally done for 14 days. This probably originates in the pharmacopoeial sterility tests where none of the major pharmacopoeias have for many decades asked for any longer incubation period. The exception to this is the Australian regulatory agency (the Therapeutic Goods Agency, TGA) which asks for 21 days' incubation to comply with its sterility test. Studies by the TGA have been influential in driving the *USP* and *PhEur* sterility test incubation period for the membrane filtration method up from seven to 14 days, even though there is no clinical evidence that the seven-day test has failed to protect the public. If the media fill is to be considered as an exhaustive search for potential viable microbial contaminants then the duration of incubation is potentially limitless. It is well known that some coryneform bacteria require 28 days or more incubation to produce visible turbidity in TSB. It is probably good conservative advice to incubate validation media fills beyond the 14-day period and justify future routine media fill incubation at 14 days or whenever the last contaminant was detected in the extended validation exercise, whichever is the longer.

There has been some controversy over the temperature of incubation for media fills — 20–25°C or 30–35°C. Any choice will always be open to criticism. Both temperature ranges (and probably some others, too) can be reasonably justified. Incubation at both temperatures is widely used, but this still leaves the decision over which temperature should be used in the first seven days, and which in the last seven days of incubation (or indeed should there be another pattern?). Once again both options are justifiable, and neither is worth an acrimonious argument with a regulatory inspector. One can only hope that facilities subject to inspection by different national agencies do not encounter single-minded inspectors with differing outlooks.

It is usual to incubate the filled units for seven days in their normal orientation, and for seven days upside down. The principle is to ensure that all of the internal surfaces of the container and closure are bathed in media for long enough to allow any adherent contaminants to be resuscitated, recover and grow. Almost always incubation in the correct orientation takes place over the first seven days, and upside-down incubation in the second period of seven days. The opposite approach could be as well justified.

The amount of media filling each container should be sufficient to reach halfway up the height of the container so that every internal surface is bathed by the medium for at least seven days. This is not always done. This factor should be taken into account when determining how the media fill is conducted.

It is advantageous to know if there are any contaminants in the media fill as soon as possible. Visual inspection, without disturbing the units, is normal on a daily or every-second-day basis. A thorough visual inspection should be conducted at seven days when the units are inverted, and 14 days when incubation is complete. Damaged or cracked units may be excluded from the results. The total number of units checked at the end of incubation, plus any removed for reasons of damage, should reconcile exactly with the numbers filled and presented for incubation. Reconciliation limits such as plus or minus 5% used in other aspects of pharmaceutical manufacture are unacceptable.

Visual inspection should be undertaken in good daylight or artificial light by personnel with good eyesight. These personnel should be subject to periodic sight tests. Turbidity is the typical indication of microbiological growth, but personnel assigned to this task should also be alert to the possibility of pellicle formation on the surface of liquid media and other forms of microbial growth. Visual inspection becomes more difficult with tinted glass containers. It is certainly most awkward for ophthalmic ointments where the contents have to be squeezed out (usually on to white paper), and examined for growth as indicated by the red colouration produced from the oxidation of tetrazolium chloride, or by the presence of bubbles.

The microorganisms from every contaminated unit obtained in any media fill should be subcultured, purified and identified to species level. Where possible the tray number and time of filling of every contaminated unit should be retained. The identity of any microbial contaminants is a major part of the information content of the media fill. Where possible the identified microorganisms should be related to the events happening at the time when the contaminated unit was filled. This view appears to contradict the apparent obsessiveness of many pharmaceutical manufacturers, microbiologists, regulators and standards writers, to place the emphasis of contaminated media fills on the numbers of contaminated units, or on the proportion of contaminated to uncontaminated units. There is practically no information content in knowing that there were two contaminated units in a media fill of, say, 4000 units. Conversely, knowing that the two contaminants were, for example, pseudomonads or micrococci, points the experienced microbiologist to the most likely source of contamination and allows intelligent diagnosis of the problem and focused corrective or preventive actions.

7 MEDIA FILLS: APPLICATIONS

Media fills are used in validation of aseptic processes as one of the final stages of performance qualification. They are also repeated periodically in routine operation of aseptic processes. It is arguable whether this latter application should be categorized as part of validation review or as part of environmental monitoring. Either way the outcome is the same — the media fill is a method of gathering information about microbiological contamination.

7.1 Media Fills in Validation of Aseptic Processes

The *Guideline on Sterile Drug Products Produced by Aseptic Processing* (FDA, 1987) refers to media fills as an "acceptable method of validating the aseptic assembly process." By 1994, the *Guideline to Industry for the Submission Documentation for Sterilization Process Validation in Applications for Human and Veterinary Drug Products* (FDA, 1994) said that specifications for media fills *should* be among the information submitted in support of sterility assurance for products manufactured by aseptic processing.

In the U.K., the 1983 "Orange" Guide (Department of Health and Social Security, 1983) gave media fills as *an example* (albeit the only example provided) of how the "efficacy of aseptic procedures should be validated." This has been succeeded by the 1992, 1997 and 2003 editions of the Commission of the European Communities' *Good Manufacturing Practice for Medicinal Product*s (CEC, 1992, 1997, 2003) which state that "validation of aseptic processing *should* include simulating the process using a nutrient medium." *Should* is a strongly directive verb in the language of these requirements.

In the last ten years media fills have, in the eyes of the regulatory bodies, developed from a reasonably good way of validating aseptic processes, through to the preferred way of validating aseptic processes, to an essential requirement of a properly validated aseptic process. It is now highly unlikely that any regulatory submission for a new aseptically filled sterile pharmaceutical product would be acceptable without supportive media fill data. It is also unlikely that a manufacturer of an existing aseptically filled sterile product would escape severe regulatory criticism if media fill data were unavailable.

It is now well accepted in the pharmaceutical manufacturing industry that validation is an exercise intended to confirm that a process is capable of operating consistently. As far as asepsis is concerned, the consistency of the contamination control "engineering" of a process is qualified by three successive replicate media fills done on separate days. Completion of the media fills is usually the factor that dictates the time of handover of the process for routine usage.

New aseptic processes require validation by media fill. Any process (irrespective of the equipment being old or new) beginning in a new clean room requires media fills as part of validation. A new filling machine in an established clean room requires validation media fills.

The trickier decisions arise over container sizes. It is quite probable that a range of container sizes may be filled on the same filling line. The question then arises over the necessity to perform media fills on all sizes, and in validation in particular, whether it is necessary to replicate each size through three media fills. The glib answer is that media fills should only be necessary for the container size that takes longest to fill and has the widest neck diameter. This combination presents the greatest potential for contamination and therefore addresses the contamination potential for all smaller sizes. However, this is not necessarily true. Wide-neck containers may be more stable than narrow-neck containers. Therefore the wide-neck filling process may be arguably less susceptible to contamination because there are fewer personnel intrusions necessary for rectifying fallen containers. Glass moulders often use a common neck or flange mold for different capacity vials; it would be usual for vials with capacities from 10 ml to 100 ml to have identical necks and flanges. There is probably no sensible way of rationalizing media fills to fewer than two container sizes on a multicontainer filling line. The decision over what and how many sizes to include in a media fill validation protocol is judgmental. For regulatory purposes the reasons for taking particular decisions must be justified and documented. If the rationale for performing media fills on more than one container size is based on the risks of contamination arising from a different source, or different balance of sources, rather than from a scale-up of risks from the same source, then it is logical that the three replicate media fills thought necessary to verify consistency of control must be performed on each container size.

The initial significant formality of validating media fills is the protocol, based on three principles.

1. **The first principle** of the protocol is that the process that is to be validated has to have been already defined and documented. In other words, draft operating SOPs have been prepared and personnel have been trained in them.
2. **The second principle** is that the test method, in this case the media fill, has been defined and documented. Importantly the protocol must define the number of units that are to be filled.

   The minimum number of units expected is 3000. The origins of this figure are worth justifying. In principle, it is an expression of the minimum number of units for which a contamination rate of no more than one contaminated unit in 1000 units (0.1%) can be demonstrated with 95% confidence. But, why a contamination rate of no more than one contaminated unit in 1000 units (0.1%)? And why with 95% confidence?

   In 1971, Tallentire and co-authors wrote that "sterility testing has several serious defects, not least amongst them being the high frequency of spurious results, sometimes called "false positives," due to contamination during testing. When measured using a population of items known to be sterile under best known test conditions, this frequency is approximately 1 in 103" (Tallentire et al., 1971).

   The view that processes involving aseptic manipulation are limited by test-related contamination at or around a frequency of 1 in 1000 originate in this 1971 paper. The one in 1000 level also ties in with the regulatory expectation of sterility test failures within any particular laboratory as no greater than 0.5% of all tests conducted.

   This assertion is based on the typical sterility test involving aseptic transfer from 20 product units; therefore 0.5% test failure represents aseptic transfer from 1000 units.

The contention was based on the technology of the 1970s. Asepsis has moved on considerably since then but the one in 1000 limit has become attached to media fills, probably because it is a practical benchmark for the number of units that can be filled or incubated, etc.

The PDA (PDA, 1981) supported a limit of no more than 0.1% contamination for media fills in its 1981 monograph (an unofficial voluntary standard). It added that this should be demonstrated with 95% confidence and that at least 3000 filled units are required to achieve this. No reason for choosing 95% confidence rather than 99% confidence or 90% confidence is given.

The idea of 3000 units and 95% confidence reappeared in the FDA 1987 Guide and has become part of the regulatory industry and expectation of media fills.

The association of 3000 units with 95% confidence of assuring a contamination rate of no more than 0.1% has been elaborated by Halls (1994), supported from two different mathematical positions.

- The PDA (1981) references the following equation of an "operating characteristic" curve to describe the probability of detecting one or more contaminated items in a sample size N taken from a population with a contamination rate of 0.1%:

$$P_{(x<0)} = 1 - e^{-NP}.$$

When $P_{(x>0)}$ is made equal to 95%, this equation describes how large a sample size, N, needs to be taken from a universe in which there is 0.1% of contaminated units to find at least one contaminated unit on at least 95% of occasions when samples are taken. In practice, 95% confidence cannot be achieved with a sample size of less than 2996.

Alternatively, the measured contamination rate in a media fill may be regarded as an estimate of the true contamination rate (P) in the underlying population that may be higher or lower than the measured rate (P_{est}). The reliability with which P_{est} can be claimed to be a true reflection of P can be calculated from the confidence limits of P_{est}.

The 95% confidence limits around P_{est} may be calculated from the expression,

$$P_{est} - hP_{est}Q_{est}/N < P < P_{est} + P_{est}Q_{est}/N,$$

where h is the number of standard deviations appropriate to particular confidence limits (1.96 for 95% confidence).

If 0.01% is regarded as the upper 95% confidence limit of the lowest measurable number of contaminants obtainable in a media fill (one contaminated unit) the lowest value of N can be calculated to be close to 3000 units.

The number of units (if any) in excess of the 3000 required to be filled is a further important decision, and there are several views on how it should correctly be made.

- One view holds that the number of units filled should be related to the product batch size. This is difficult to reconcile with the fact that the media fill is a process test and should not logically be related to product batching. Different products filled into the same containers on the same filling machine could easily have different batch sizes, perhaps dictated by some complexity of compounding. To which of these batch sizes should the number of media fill units be related? If this approach is taken, the pragmatic answer is usually the largest of the batch sizes. Guidance on media fill dimensions in relation to product batch size given in ISO/IS 13408 *Aseptic Processing of Health Care Products* (ISO, 1997) is summarized as Table 3.3. In practical terms this guidance applies only to small batch sizes; for normal production batch sizes ISO supports only a minimum media fill size of 3000 units.

Table 3.3 Minimum Numbers of Media Fill Units Related to Production Batch Size from ISO/IS 13408 *Aseptic Processing of Health Care Products (ISO, 1997)*

Number of units in production batch	Minimum number of units for validation media fills	Minimum number of units for periodic media fills
< 500	5000 in ten or more runs	Maximum batch size per run
≥ 500–2999	5000 in three or more runs	Maximum batch size per run
≥ 3000	9000 in three runs	3000 per run

- A second view is that the media fill should be run over the same time as an operating shift. In many cases this amount of elapsed time may be necessary to simulate all of the potential contaminating events arising in a process. In other cases, e.g., with high-speed ampoule filling lines, it could result in vast numbers of units being filled. As long as there are no contaminated units present, this approach to filling ampoules gives good assurance of asepsis. Its logic breaks down when contaminated units are identified; perhaps three or four contaminated units would be insignificant in comparison to the overall large numbers filled. Regardless, they may be significant to contaminating events that occurred during filling but the effect of the large dimensions of the media fill is to dilute their impact.
- The third view, and the one supported by the author, is that the dimension of the media fill should be dictated by the time necessary to allow simulation of all of the potential contaminating events. The identification

of *all* potential contaminating events is in the long run a matter of opinion. Nonetheless, there are techniques such as failures modes and effects analysis that can be used to create a documented structure around the development of these opinions. This type of approach adds to the knowledge of the process if done properly. If done in a cursory manner it is open to abuse.

3. **The third principle** of the protocol is that acceptance criteria must be predetermined. In the case of media fills a maximum number of contaminated units must be specified for the media fill, and indeed the underlying aseptic process, to be acceptable. If the acceptance number is exceeded in any one of the three validation media fills, appropriate action must be taken and the media fill(s) repeated until three successive successful media fills are obtained. Ideally the appropriate action is preventive, i.e., action appropriate to preventing a further recurrence should be taken, probably involving some change in working practice and to the operating SOP.

In the real world the action is most often corrective — something like a re-disinfection of the filling room or retraining of personnel. This is because it is not usually easy to accurately diagnose the source of contamination in a media fill, and this difficulty is greatest for a new process (and one would expect validation to be done for a new process in its broadest sense).

The question arising out of the predetermination of acceptance criteria is exactly how many contaminated units are tolerable? This is not an easy question to answer.

As a starting point, if we take the statistic of no more than one contaminated unit in 1000 as the acceptance limit, and 3000 units as the minimum number of units in a media fill, then we might reasonably expect that zero, one, two or three contaminated units in 3000 would be acceptable, and four or more contaminated units unacceptable. Up to four contaminated units would be acceptable in 4000, up to five in 5000, etc.

This approach was overtaken by the PDA recommendation (1981) that the limit of no more than one contaminated unit in 1000 should be met with 95% confidence. In relation to 3000 units filled, compliance with this modification to the one-in-1000 limit would only be acceptable with zero or one contaminated units.

Slightly different (but in practical terms insignificant) mathematical treatments result in recommendations of "pass zero, fail one or more contaminated units in 3000" or "pass one or fewer contaminated units, fail two or more contaminated units in 3000."

When media fills require numbers of units larger than 3000 it might be considered reasonable to increase the number of contaminated units permissible beyond zero or one. The guidance in ISO/IS 13408 allows maximum numbers of contaminated units ranging from one in a 3000 unit media fill to 11 in a 17,000-unit (approximately) media fill (ISO, 1997).

Bernuzzi et al. (1997) examined these recommended limits and concluded that they became weaker as the total number of filled units increased. (One positive in a 5000-unit media fill does not have the same meaning as 10 units positive when 16,970 units are filled.)

These authors attempted to develop an alternative set of limits for media fills, but in all cases found the same statistical frailty as numbers of units filled increased. The limits from ISO/IS 13408 (ISO, 1997) and from the most rigorous plan of Bernuzzi et al. (1997) are summarised in Table 3.4.

Table 3.4 Maximum Permissible Numbers of Contaminated Units in Media Fill According to ISO/IS 13408 *Aseptic Processing of Health Care Products* (ISO, 1997) and to the More Rigorous Scheme of Bernuzzi et al. (1997)

Number of units filled	Maximum permissible number of contaminated units	
	ISO/IS 13408 (1997)	Bernuzzi et al. (1997)
3000	1	0
4750	2	–
6300	3	–
7200	–	1
7760	4	–
9160	5	–
10520	6	–
11500	–	2
11850	7	–
13150	8	–
14440	9	–
15710	10	–
15800	–	3
16970	11	–
20200	–	4

It is, however, all very well in principle and in statistics to present limits such as these. In practice it is unrealistic that, for example, a manufacturer of aseptically filled ampoules would repeatedly tolerate (or be allowed by the regulatory agencies to tolerate) six contaminated units in media fills of 10,000 units as these recommendations appear to suggest. It is also unrealistic that a manufacturer of blow-fill-sealed ampoules would repeatedly tolerate even four contaminated units in media fills of 20,000 units.

In practice any number of contaminated units in excess of zero or one would have to be investigated seriously by any conscientious pharmaceutical manufacturer, irrespective of the overall dimensions of the media fill. This is particularly true in validation. It is outside the experience and belief of the author that any ethical pharmaceutical manufacturer would approve validation of an aseptic process in which three or more contaminated units appeared in validation media fills.

Contaminated units are the stimulus for process improvement. The practical limit in all media fills is that there should be no more than one, possibly (in very unusual circumstances) two contaminated units. Larger numbers of contaminated units must elicit preventive action and improved control.

In summary, it makes the best sense that validation media fills should be composed of a number of units in excess of 3000 sufficient to allow for enough elapsed time to simulate all predicted potential contaminating events, and no more than one contaminated unit should be allowed in any single run no matter how many units are filled in total.

7.2 Periodic Media Fills in Routine Operation

It is unlikely that any responsible regulatory body would tolerate a frequency of less than twice a year for periodic media fills. Media fills are probably the most sensitive method of detecting unexpected sources of process contamination. The regulatory standpoint coming from the principle of patient protection is that if unexpected process contamination occurs in a media fill, and is considered sufficient to compromise the sterility of past product, they would expect market withdrawal. Following this logic, the greater the frequency of periodic media fill, the lower the risk to the patient, and the lower the commercial risk to the manufacturer.

Media fills are generally done on every filling line at least twice a year. (88.5% of the respondents to the PDA's 1996 survey performed media fills at least twice a year.) Within this program it is sensible to ensure that on multicontainer filling lines every container size has been filled at least once in a reasonable time frame, say, over two years. Otherwise the possibility of unexpected contamination as it relates to a particular size may never be addressed.

It is also arguable that at least one of those sizes identified in the validation protocols as presenting the worst risks of contamination should be tested on every occasion of periodic media fills. This is usually achieved by setting up some sort of matrix approach to periodic media fills on multicontainer filling lines. Of course it is much easier on a single-container size, single-volume filling line.

Periodic media fills should be done at the end of a routine production operation. Care should be taken to run a few litres of sterile water through the filling setup, to flush out any product-related inhibitory substances, before filling the placebo. This is intended to address the two possibilities of contamination buildup in a filling room over a period of manned operation between clean-ups, and of lapse of operator discipline as a result of tiredness.

The exception to this is for aseptic filling of sterile antibiotics. Here the filling room must be cleaned up and all antibiotic traces removed before the placebo is brought in and filled. This ensures that recovery of contaminants is not inhibited. This reinforces the argument concerning the purpose of media fills as related to *process* contamination, rather than to *product* sterility. The media fill is intended to disclose process contamination, regardless of whether the contaminants would survive or die in the product.

ISO/IS 13408 (1997) recommends a two-tier approach of alert and action to limits for periodic media fills. It does not, however, elaborate on how the two levels should be applied. The action limits are listed under ISO/IS 13408 in Table 3.4. The alert limits are lower.

The divergence between the ISO/IS 13408 action limits and acceptable reality has already been discussed in relation to validation media fills. Reality for a valid aseptic process that has been transferred to routine control is that there will have been no more than one contaminated unit per media fill run, irrespective of the total number of units filled.

It is axiomatic that the periodic media fill should not generate significantly worse results than the validation media fill without some appropriate action being taken. Here it is suggested that limits for periodic media fills should be related to the results obtained in validation media fills as summarised in Table 3.5.

Table 3.5 Recommended Action Limits for Periodic Media Fills Irrespective of Total Numbers of Units Filled and Related to Results from Validation Media Fills

Numbers of contaminated units actually occurring in three successive validation media fills	Action limit (numbers of contaminated units) for marginal failures in periodic media fills	Action limit (numbers of contaminated units) for consequential failures in periodic media fills
0, 0, 0	≥ 1 but < 3	≥ 3
0, 0, 1	≥ 1 but < 3	≥ 3
0, 1, 1	≥ 2 but < 4	≥ 4
1, 1, 1	≥ 2 but < 4	≥ 4

This may appear a little radical but it is no more than common sense, and reflects what is done in many other less critical industries, than those manufacturing sterile pharmaceutical products. Fundamentally, this suggestion proposes that a new aseptic process should be developed such that there is minimal evidence of contamination and this is surely what is being done already. Then the response limits for routine media fills should be based on the process capability demonstrated in validation.

The numbers of permissible contaminated units shown in Table 3.5 are presented in a two-tier approach for which both levels demand action. The difference in the approach is in the consequences of the actions to production, to scheduling and to past product.

The term "action limits for marginal failures" is used here rather than "alert" limit, because the author's opinion is that all contaminants found in media fills merit some action, and that the use of the "alert" term detracts from this.

The action limits for marginal failures allow for the extremes of statistical variation from the validation media fill results that might be expected, without any significant change in the real contamination rate. In the author's experience, media fill contaminants are fairly rarely found in well-controlled facilities, but when they occur, most fall into this marginal category.

Identification and investigation are essential. The possibility that they may not be a statistical phenomenon should not be discounted. *Bacillus* spp. should, for instance, be treated with extreme suspicion in relation to the possibility of some systematic problem with nonsporicidal disinfection, or of residual air in autoclave loads, etc.

Actions from marginal failures, which do not appear to have arisen from a systematic failure of one of the systems necessary for the maintenance of asepsis, are best dealt with by counselling, retraining, and improved supervision of operators. The media fill should be repeated as soon as possible. A further media fill on the container size implicated should be scheduled into the next periodic media fill, in addition to those sizes defined by the predetermined matrix.

Successive marginal failures on the same container size should be treated as a consequential failure, as also should marginal failures on three or more successive media fills on the same filling line, irrespective of container size. Other circumstances of repeated failures within the marginal range may also be indicative of process conditions that have deteriorated from the validated condition, and should be treated as infringements of the action limits.

Table 3.5 gives action limits that are described as consequential. These limits are well beyond the expected variation seen in the validation media fills and must therefore be interpreted as indicators of real loss, or genuinely deteriorating control levels. It is reasonable to expect that the potential for any patient risk should be minimized while these failures are being resolved. Product manufactured on the filling line after the date of the media fill, and product still in the company's warehouses, should be quarantined until the failure investigation is completed.

Ideally production on the line in question should be suspended pending the outcome of the investigation. In practice it may be advantageous to the investigation for production to continue, but this decision should not be taken lightly in view of the commercial risk of possibly having to reject the product made in that period.

The most important factor in the failure investigation is the identification of the contaminants. Any microbiologist should be able to categorise identified contaminants within their most likely sources to the environment (air, dust, etc.), water, or human sources. An experienced QA microbiologist may be able to pinpoint the contaminants to their origins in the facility (e.g., nonsterile disinfectants, water leakage, worn-out garments, etc.) or to general weaknesses in control.

Environmental microbiology is not an exact science. A weakness in control will always be a systematic weakness even though it may not manifest itself every time it is tested by media fill. If a specific problem is diagnosed it should be traced, if possible, to the time it began.

The identified contaminants should be considered for their ability to survive in the products filled on the line in which the consequential media fill occurred. The importance of this information is to determine if sterility has been compromised. In a multiproduct filling line the decision might be different for different products. If sterility is compromised then product must be withdrawn from the market.

Once the failure investigation is complete and corrective or preventive action implemented, it is customary to repeat the media fill. Some would argue for revalidation of the line by repeated media fill, but this decision should be contingent upon the extent of the corrective or preventive action implemented.

In some instances of consequential failure it may not be possible to pinpoint the cause and corrective or preventive action cannot therefore be targeted. In such cases it is normal to clean, disinfect, fumigate, counsel, train and improve supervision overall. Three repeat media fills should be done to counterbalance the uncertainty of the diagnosis.

Where repeat media fills have to be done, the date of recommencement of production is a business that requires account to be taken of the uncertainty of the diagnosis of the cause of the problem. Where the source of media fill failure is quite clear, and preventive action self-evident, it is probably a reasonable risk to recommence production before the media fill incubation is complete. The commercial risk is greater where diagnosis of the problem is unclear.

The major practical issue of periodic media fills is how to respond to the results. This is less problematic (in principle, if not always in practice where deadlines have to be met) in validation than in periodic media fills. The major issues are:

- Should production on a particular filling line be allowed to continue if media fill results are unfavourable?
- Media fill results are not available until 14 days after the media fill has been conducted. What should be done with the product manufactured between these dates when results are unfavourable?
- Media fills are only done every six months. What should be done with the product manufactured since the last successful media fill when results are unfavorable?

If a company has the luxury of running terminally sterilized products on the same filling line as aseptically filled products, there is the opportunity to run the line and investigate a media fill failure while fully operational. If only aseptically filled products are filled on the line, filling should be suspended until investigations are complete and repeat media fills are satisfactory. It is exceedingly difficult to reach a firm conclusion unless the line is running. Although running a series of repeat media fills is essential, the hope is usually that they will pass rather than providing information.

It is normally recommended to "freeze" all of the product still in company control aseptically filled on a failed media fill line until investigations are complete and repeat media fills have given the line the "go ahead." This strategy, although fine in principle, usually raises significant pressure from marketing and distribution over stock-outs or impending stock-outs.

If a media fill fails and this is traceable to a failure in one of the component systems making up the sterility assurance system, there is little choice but to reject and recall back to that date, unless the regulatory bodies can be convinced otherwise. A responsible recall initiative from a company is generally less harmful than a recall requested by an inspector who discovers the matter later on. An example of this could be a tear in a HEPA filter — recall back to the last satisfactory *in situ* integrity test.

The outcome of the investigation of most marginal media fill failures is inconclusive, often from some human commensal microorganism shed by an operator, not necessarily on point-of-fill, possibly even when unloading stoppers from an autoclave. It would not be sensible to recall for this type of phenomenon, and in mitigation there could be some work done on the potential for the particular microorganism to survive and grow in specific products filled on that line.

REFERENCES

Bernuzzi, M., Halls, N.A., Raggi, P. Application of statistical models to action limits for media fill trials. *European Journal of Parenteral Sciences*, 2: 3–11, 1997.

Commission of the European Communities (CEC). *The Rules Governing Medicinal Products in the European Community. Volume IV. Good Manufacturing Practice for Medicinal Products*. Luxembourg: Office for Official Publications of the European Community, 1992, 1997, 2002.

Department of Health and Social Security. *Guide to Good Pharmaceutical Manufacturing Practice* (the "Orange Guide"). London: Her Majesty's Stationery Office, 1983.

Food and Drug Administration of the United States Department of Health and Human Services (FDA). *Guideline on Sterile Drug Products Produced by Aseptic Processing*. Rockville, MD: Center for Drugs and Biologics, 1987.

Food and Drug Administration of the United States Department of Health and Human Services (FDA). *Guideline to Industry for the Submission Documentation for Sterilization Process Validation in Applications for Human and Veterinary Drug Products*. Rockville, MD: Center for Drug Evaluation Research and Center for Veterinary Medicine, 1994..

Halls, A. *Achieving Sterility in Medical and Pharmaceutical Products*. New York: Marcel Dekker, 1994.

International Organization for Standardization (ISO). ISO 13408-1 *Aseptic Processing of Health Care Products — Part 1: General Requirements*, 1998.

Medicines Control Agency (MCA). *Rules and Guidance for Pharmaceutical Manufacturers and Distributors, 1997*. London: The Stationery Office, 1997.

Parenteral Drug Association (PDA). *Validation of Aseptic Filling for Solution Drug Products, Technical Monograph No 2*. Bethesda, MD: Parenteral Drug Association Inc., 1981.

Parenteral Drug Association (PDA). Technical Report No. 24. Current practices in the validation of aseptic processing. *PDA Journal of Pharmaceutical Science and Technology*, 51: Supplement S2, 1996.

Tallentire, A., Dwyer, J., Ley, F.J. Microbiological quality control of sterilized products: evaluation of a model relating frequency of contaminated items with increasing radiation treatment. *Journal of Applied Bacteriology*, 34: 521–534, 1971.

Chapter 4

Contamination of Aqueous-Based Nonsterile Pharmaceuticals

Nigel Halls

CONTENTS

INTRODUCTION

Pharmaceutical preparations are expected to be efficacious, safe and affordable.

0-849-32300-2/04/$0.00+$1.50

The presence of microorganisms in pharmaceutical preparations may reduce their efficaciousness and make them unsafe for the patient.

The severity of the consequences of microorganisms present in pharmaceutical preparations differs according to the purpose of the preparation and its route of administration. Some preparations for instance, must be free from all viable microorganisms (sterile preparations); others are not required to be sterile, but are subject to certain restrictions on the number and types of tolerable microorganisms to ensure their efficaciousness and safety.

This chapter is concerned with the microbiological control of aqueous-based nonsterile preparations — topical (lotions, creams, gels, ointments), aqueous oral (solutions, suspensions, syrups) and aqueous inhalation preparations — because they are more susceptible to microbiological problems than other nonsterile dosage forms.

1 SUSCEPTIBILITY OF PHARMACEUTICAL PREPARATIONS TO MICROBIOLOGICAL PROBLEMS

Two problem, broadly speaking, are created by the presence of microorganisms in pharmaceutical preparations. First, they may harm the patient by causing infection. Second, they may alter the composition of the preparation to the extent that it may not function in the way it was intended, or that the patient may reject it as "spoiled."

Generally (but not invariably), high numbers of microorganisms are necessary for either problem to arise in, or from, a pharmaceutical preparation. Two processes must occur to allow microbial contaminants to reach problematic numbers:

- There must be an initial contamination of the preparation by microorganisms
- The microbial contaminants must proliferate in the preparation, and must metabolise, grow and multiply

The probability of any initial contamination arising is a reflection of how well good manufacturing practices (GMPs) have been applied. However, the likelihood of proliferation is largely a function of the composition of the preparation itself. There are three alternative fates for microorganisms contaminating pharmaceutical preparations (Figure 4.1):

- They may die
- They may survive without proliferating
- They may metabolize, grow and multiply

(The fourth remote possibility is that microorganisms might "escape.")

Although microbiology is not a very precise science, it can quite confidently predict the most likely fate of contaminants in various types of pharmaceutical preparation. Microorganisms require water to metabolize, grow and multiply.

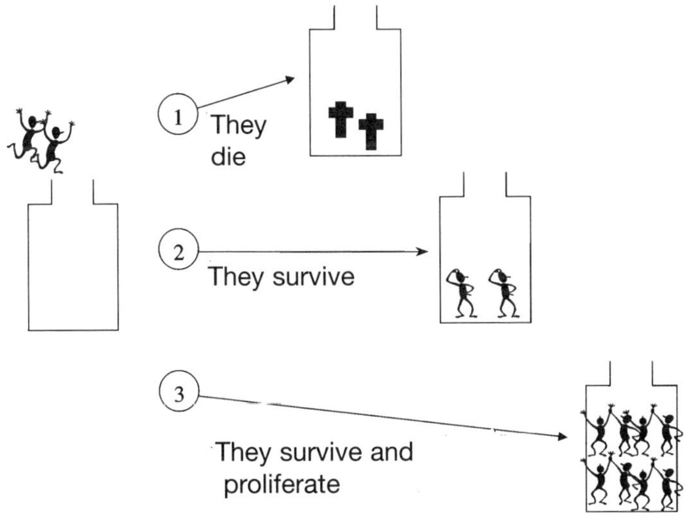

Figure 4.1. Alternative fates of microbiological contaminants.

Strictly speaking they require "free" or "unbound" water as measurable through water activity.[1,4,8] There is only a very low probability of microbial proliferation in nonaqueous pharmaceutical preparations. Their most likely fate is death through desiccation — at worst, desiccation-resistant types (e.g., *Bacillus*, *Micrococcus*) survive without multiplying. The greatest opportunity areas for microbiological proliferation are in *aqueous-based pharmaceutical preparations*.

Of all aqueous-based preparations, inhalations, oral solutions and suspensions, topical lotions and creams are most susceptible to microbiological problems (Table 4.1). Syrups and topical gels are of medium susceptibility. Ointments are of quite low susceptibility to microbiological problems. Correspondingly these differences in susceptibility to microbiological problems define the attention that should be given to addressing microbiological concerns in formulation, in the choice of starting materials and in the control of manufacture.

Table 4.1. Susceptibility (Based on Water Content) of Pharmaceutical Preparations to Microbiological Problems

Low Susceptibility	Medium Susceptibility	High Susceptibility
Ointments	Oral syrups	Aqueous inhalations
	Topical gels	Oral solutions
		Oral suspensions
		Topical lotions
		Topical creams

2 MICROBIOLOGICAL CONTAMINATION LIMITS IN PHARMACEUTICAL PREPARATIONS

Ideally all pharmaceutical preparations would be completely free from any form of contamination, including microbiological contamination. However, this is not truly necessary and is impractical for most preparations. This is recognised by the pharmacopoeias. Both the *United States Pharmacopeia* (*USP*) and the *European Pharmacopoeia* (*PhEur*) contain chapters describing testing of pharmaceutical preparations for compliance with microbiological contamination limits. The *USP* specifies mandatory microbiological limits within most of its guides for pharmaceutical preparations. *PhEur* specifies nonmandatory microbiological limits for general categories of pharmaceutical preparations.

2.1 The Microbial Limit Tests

Methods suited to testing products for compliance with microbiological contamination limits for pharmaceutical preparations are given in Chapter 6 of *USP* and Chapter V.2.1.8 of *PhEur*. Although differences between the two methods are often stressed by microbiologists concerned with pharmacopoeial harmonization, they are essentially minor. The two pharmacopoeias follow quite similar principles.

2.2 Counts for Compliance

There is a choice of four approaches to compliance with quantitative limits:

- Pour plates
- Membrane filtration
- Surface spread plates
- The "most probable number" approach

Pour plates and membrane filtration are widely used; surface spread plates are more commonly used for counting microorganisms in pure culture work. The most probable number approach is a method of last resort and its supporting statistical tables are limited, as they do not properly address mold recovery. Each of these four approaches splits into two methods.

- Counting numbers of aerobic bacteria per unit weight or volume of product
- Counting numbers of yeasts and moulds per unit weight or volume of product

Microbiologists debate endlessly whether the two methods are mutually exclusive, and how to derive a total count if they are not. Most of this debate is only of academic interest. For quality control (QC) purposes safe conservative assumptions are that:

- Bacteria are never recovered in the test for yeasts and molds
- Yeasts and molds are never recovered in the test for aerobic bacteria
- The total count is the sum of the two

With properly formulated preparations manufactured under controlled GMP conditions, actual recoverable numbers are rarely within a decimal order of magnitude of the specified limits. *Any* manufacturer producing a nonsterile aqueous-based preparation close to (or frequently failing) the pharmacopoeial limits for numbers of microorganisms, is already in trouble!

2.3 Tests for Absence of Specific Indicator Microorganisms

Both pharmacopoeias describe methods for determining if particular microorganisms are present or absent from specified weights or volumes (10g or 10 ml in *USP*, 1 g or 1 ml in *PhEur*) of pharmaceutical preparations. These methods involve incubation of the pharmaceutical preparation in enrichment media, which encourages growth of one particular microorganism at the expense of others. This is followed by surface spread plating on media (selective and differential) upon which the particular microorganism takes a distinctive and easily recognizable colonial appearance.

The pharmacopoeias specify methods applying to *Staphylococcus aureus*, *Pseudomonas aeruginosa*, *E. coli* and *Salmonella* spp., with a method for each microorganism. Within each method there are options around choice of media and, in some cases, incubation temperatures. *PhEur* includes a broader category for "enterobacteria and certain other Gram-negative microorganisms" and recommends its application to Category 2 products (topical and respiratory preparations and transdermal patches).

Microorganisms for which methods are described in the pharmacopoeias are pathogenic. However, they have not been included solely for their pathogenicity: they are "indicator" or "index" microorganisms, chosen because they are easily recoverable and recognizable using robust and readily available methodology.

They are indicators of excessively contaminated raw materials and unhygienic manufacturing practices, providing a risk index for the presence of low numbers of other pathogens, *for which feasible methods of detection are not available.*

It is out of the question that batches of routinely produced pharmaceutical preparations should be critically examined and certified free from all pathogens and potential pathogens; quite simply, nothing would ever be released within its shelf life. However it would be extremely naïve to believe that absence of all four selected indicator microorganisms is synonymous with absence of all pathogens and potential pathogens.

A technical curiosity of the methods, media and conditions recommended in the pharmacopoeias for recovery of these selected indicator microorganisms is that they derive from medical microbiology. As such, the media were originally designed to encourage the growth of a particular type of microorganism that would be most likely present as a minor component of a much larger microbial population in pathology samples, faeces, etc.

In pharmaceutical preparations, the selected indicator microorganisms would most likely be accompanied by very few other types, if present at all: *Salmonella,* for instance, has no recent history of ever being isolated from properly manufactured and formulated pharmaceutical preparations. There are substantial data to show that under these circumstances, standard general-purpose media are just as effective for the isolation of these selected indicator microorganisms as the complicated enrichment and selective media recommended.[3]

2.4 International Pharmacopoeial Guide Limits

In setting the limits for guides appropriate to particular pharmaceutical preparations, both *USP* and *PhEur* take account of

- The significance of microorganisms to different types of product
- The way in which the product is used
- The potential hazard to the patient.

Although the pharmacopoeias describe tests for counting two different groups of microorganisms, and tests for detecting the presence or absence of four different selected indicator microorganisms, there is only one single guide in *USP XXVI* to which all these restrictions apply (Silver Sulfadiazine Cream).

The *USP XXVI* has 18 guides for oral liquids containing microbiological specifications, 40 for creams and lotions, 17 for ointments and gels, and only three for aqueous inhalations. Table 4.2 makes it clear that "typical" *USP* microbiological limits applying to oral liquids are for quantitative limits and restrictions on the absence of *E. coli* and *Salmonella* spp. "Typical" *USP* microbiological limits applying to topical preparations are for absence of *Staphylococcus aureus* and *Pseudomonas aeruginosa*.

The distinction between the two groups of limits for pharmaceutical preparations is that enteropathogenic microorganisms such as *E. coli*, *Salmonella* and other similar types are infective through ingestion in the way that oral liquids are taken, but are not typically infective through the skin. *Staphylococcus aureus* and *Pseudomonas aeruginosa* and other similar types of microorganism are infective when applied to broken or damaged skin, where topical preparations are likely to be applied.

PhEur has a broadly similar approach to its categorization of products versus microbiological limits. In Section 5.1.4 it recommends (but does not mandate) limits that should be applied if necessary for broad "categories" of pharmaceutical preparation. For aqueous-based nonsterile preparations, *inhalations*, *topicals* and

Table 4.2 Microbiological Limits in *USP XXVI.*

Numbers of Guides	With/without quantitative limits	With/without limits for absence of *Staphylococcus aureus* or *P. aeruginosa*	With/without limits for absence of *E. coli* or *Salmonella* spp.
Oral solutions, Suspensions and syrups	15/3	5/13	17/1
Topical creams and lotions	2/38	40/0	6/34
Topical ointments and gels	0/17	17/0	2/15
Aqueous inhalations and nasal solutions	0/3	3/0	0/3

transdermal patches belong in category 2; oral solutions, suspensions and liquids in Category 3A.

Category 2 preparations require absence of *Staphylococcus aureus*, *Pseudomonas aeruginosa* and compliance with a quantitative limit. Category 3A preparations should comply with a quantitative limit and be free from *E. coli.*

The "typical" quantitative limit in *USP* for oral liquids is not more than 100 cfu/g or ml (aerobic bacteria). This applies to 14 of the 15 guides specifying quantitative limits in *USP XXVI.* One guide (Aciclovir Oral Suspension) has a tighter limit of not more than 10 cfu/ml, which is understandable considering that patients using this preparation are likely to have weakened immune systems. Four of the guides with limits on bacteria also have limits on yeasts and molds (not more than 10/g or ml).

PhEur applies weaker limits than *USP* to bacterial numbers in oral liquids. Category 3A, for preparations for oral administration applies a quantitative limit of not more than 10^3 aerobic bacteria/g or ml. On the other hand, *PhEur* applies a quantitative limit of not more than 10_2 fungi/g or ml to all products for oral administration — *USP* rarely applies limits to yeasts and molds.

Restrictions on particular microorganisms in *USP* are intended to apply to absence in 10 g or 10 ml. The restrictions in *PhEur* apply only to absence in 1 g or 1 ml.

2.5 Objectionable Microorganisms

It is important to recognise that pharmacopoeial limits on microbiological contamination in pharmaceutical preparations are not all embracing. Contamination limits can never adequately specify everything present in a pharmaceutical preparation that might risk infection, or lead to patient refusal. Many microorganisms are not specifically restricted from pharmaceutical preparations, but if present — even in numbers well within the quantitative limits — would clearly constitute a risk to the patient.

An example of this is *Pseudomonas cepacia*, recognized as an important human pathogen, particularly in topical products and inhalations. It would be downright irresponsible to release a preparation to market with (say) a quantitative count of 5 cfu/g versus a limit of not more than 100 cfu/g, if those colonies were identifiable with *Pseudomonas cepacia*.

The U.S. Food and Drug Administration (FDA) *Guide to Inspection of Microbiological Pharmaceutical QC Laboratories* (1998) describes *Pseudomonas cepacia* as objectionable. Clearly, "objectionability" is an important term and concept. FDA in its *Guide to Inspection of High Purity Water Systems* (1993) defines objectionable microorganisms as "organisms which can cause infections when the drug product is used as directed, or any organism capable of growth in the drug product."

The first part of this definition closely follows the approach of pharmacopoeias to evaluating the risk of infection that microorganisms can create, versus the route of administration of the pharmaceutical preparation to the patient. The second part infers that a microorganism that cannot proliferate in the product is not objectionable unless it can cause infection. In this way, some recovery of low levels of (say) *Bacillus* spp. from an oral liquid would not typically be expected to be objectionable.

On the other hand an example of an objectionable microorganism in an oral liquid could be *Gluconobacter* sp. This noninfective microorganism is occasionally found in syrups where it can proliferate and produce slimes and foul odours: its biochemical profile closely resembles that of *E. coli* but it grows poorly at 37°C, and is more likely to be recovered under conditions designed for yeasts and molds than those designed for recovery of bacteria.

Figure 4.2 shows a decision tree focusing on the issues surrounding objectionability. The first decision points are around known pathogenicity and recourse must be made to a reliable and standard text. *Bergey's Manual of Determinative Bacteriology* is reliable; the *Manual of Clinical Microbiology*[7] has better focus on pathogenicity.

The second decision point is whether the microorganism has been associated with — but is not necessarily causative of — infectious disease by the administration route, and to the target patient population. Making a decision around this issue is becoming more complicated. If only it were as simple as reaching the conclusion that in a paediatric presentation, any microorganism known to be associated with an infectious illness should be considered objectionable. The number and proportion of the immunodeficient and the immunosuppressed is increasing in the general patient population through greater longevity. (The average life expectancy of a male in the U.K. was 71.7 years in 1985 rising to 75 years in 1999 according to OECD statistics (OECD 2000.)

Increased life expectancy is anticipated for sufferers from HIV/AIDS (34,000 living in the U.K. and 900,000 living in the U.S. in 2001 (UNAIDS/WHO 2000); improved control of diabetes; the progress of transplant surgery; and more

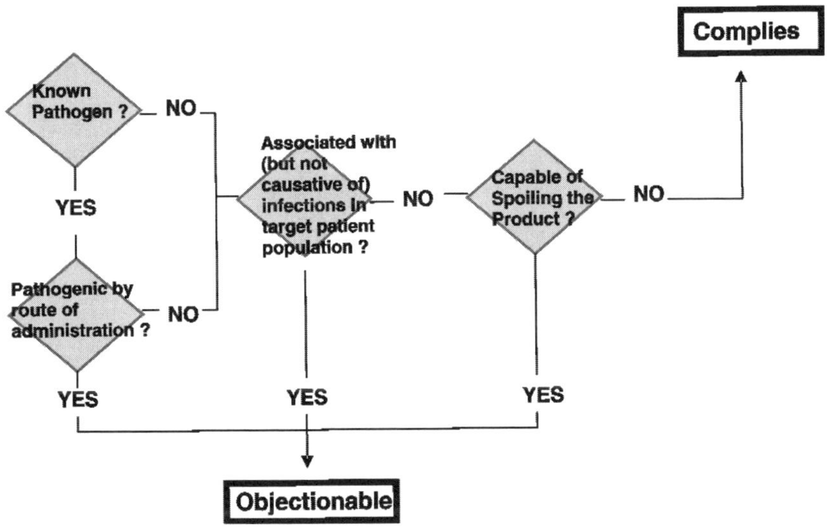

Figure 4.2. Objectionable microorganisms in nonsterile preparations.

successful applications of chemotherapy in cancer treatment. You can still suffer from indigestion, excema, asthma and the common cold while immunosuppressed and immunodeficient.

The capability of a specific type of microorganism to spoil a particular product is also problematic, as it may never have been previously addressed. A preservative efficacy test using the questionable microorganism could be the answer.

For the QA microbiologist it is important to take account of all colonies recovered in microbial limit tests and pay heed to their potential for objectionability. In the author's opinion, all Gram-negative microorganisms (particularly pseudomonads, although recognition of these has become significantly more complicated in recent years by division of the genus into numerous genera such as *Burkholderia*, *Ralstonia*, *Stenotrophomonas*, etc.) isolated from pharmaceutical preparations must be assumed to be objectionable, unless there is compelling evidence to the contrary. *Bergey's Manual of Determinative Bacteriology* and Murray's *Manual of Clinical Microbiology*[7] are reliable primary sources of information on the infectivity of microorganisms. Other information on current views on the objectionability of particular microorganisms may be obtained from regulatory publications, particularly those from FDA. For instance, a warning letter issued in September 2002 and available on the FDA Web site, indicates quite clearly that FDA considers *Serratia liquefaciens* and *Pseudomonas fluorescens/putida* to be objectionable in aqueous nasal spray products.

Table 4.3 Gram-negative Microorganisms That Should be Considered Objectionable as a Result of Infectivity

(NOTE: This list is for guidance; it is not comprehensive. Other microorganisms not listed may also be objectionable through infectivity or other reasons. Absence of any indication of infectivity in a particular range of preparations does not necessarily indicate that it is neither infective nor objectionable.)

	Topical Preparations	Liquid Oral Preparations	Inhalations
Pseudomonas (Burkholderia) pseudomallei	Causes melioidosis by contact with cut or abraded skin		Causes melioidosis by inhalation.
Pseudomonas aeruginosa	Opportunistic pathogen of man found in wound infections		Opportunistic pathogen of man found in the respiratory tract
Pseudomonas (Stenotrophomonas) maltophilia			Pathogen in immunosuppressed patients and those with cystic fibrosis
Pseudomonas (Burkholderia) cepacia	Recognised as objectionable by FDA in topical products and nasal solutions		Recognised as objectionable by FDA in topical products and nasal solutions
Pseudomonas fluorescens			Contaminated inhalation solutions recalled in U.S. in 1993
Pseudomonas putida	Contaminated barrier creams recalled in U.S. in 1998		
Acinetobacter spp.			Agents of nosocomial pneumonia
Flavobacterium (Chryseo-bacterium) meningosepticum			Cause of meningitis and pneumonia in infants
Bordetella parapertussis			Causative agent of whooping cough (transmitted in aerosols)

E. coli		Agent of gastrointestinal disease	
Shigella spp.		Causative agent of bacillary dysentery	
Plesiomonas shigelloides		Pathogenic in the human intestine	
Salmonella spp.		Causative agent of enteric fevers, gastroenteritis, etc.	
Klebsiella spp.			Associated with respiratory tract infections, pneumonia, etc.
Enterobacter spp.			Associated with respiratory tract infections, pneumonia, etc.
Edwardsiella tarda		Causative agent of a Salmonella-like enteritis	
Providencia spp.	Infective in wounds and burns		
Yersinia spp.		Causative agents of gastrointestinal infections	
Aeromonas hydrophila		Causative agent of acute diarrhoea	

Table 4.3 contains a limited list of Gram-negative microorganisms versus the types of preparation in which they should be considered objectionable on the basis of infectivity. This is not comprehensive. The fact that *Providencia* spp. is listed as infective only under topical preparations, does not necessarily mean that it may not be infective or objectionable in other types of preparations. To nonspecialists it may appear curious that some of the better known causative agents of infectious disease, e.g., *Campylobacter* spp. are not included in this list. This is because there is very little probability of them being isolated from pharmaceutical preparations by the techniques normally recommended and used. *Campylobacter* spp., for instance, require a microaerobic atmosphere (5% O_2, 10% CO_2, 85% N_2) for optimal recovery.

3 MICROBIOLOGICAL CONTAMINATION CONTROL PRINCIPLES

The principles of controlling microbiological contamination in pharmaceutical preparations are part of GMPs and are quite simple:

- Identify the sources of microbiological contamination. Where possible eliminate them. If this is not possible, minimize them
- Identify the vectors for transmitting microbiological contamination. Where possible eliminate them. If this is not possible, minimize them
- Identify critical operations and provide local protection around them
- Identify and minimize the opportunities for microorganisms to proliferate
- Monitor the effectiveness of the control measures

The practicality of controlling microbiological contamination in pharmaceutical preparations is more complicated than the theory. Some contamination risks may be generic to all facilities in which particular product types are manufactured. Other risks may be functions of particular manufacturing processes or conditions. Yet others may be functions of the products themselves (e.g., their formulations). The extent to which it is necessary to control these risks differs from one product to another because of differing risks to the patient.

3.1 Sources and Vectors for Contamination

There are several broad areas to which contamination may be traced in the manufacture of all pharmaceutical preparations. Figure 4.3 shows a schematic representation of the sources of contamination in any nonsterile manufacturing facility. The main sources are:

- Incoming raw materials and ingredient water
- Facilities, services and cleaning materials

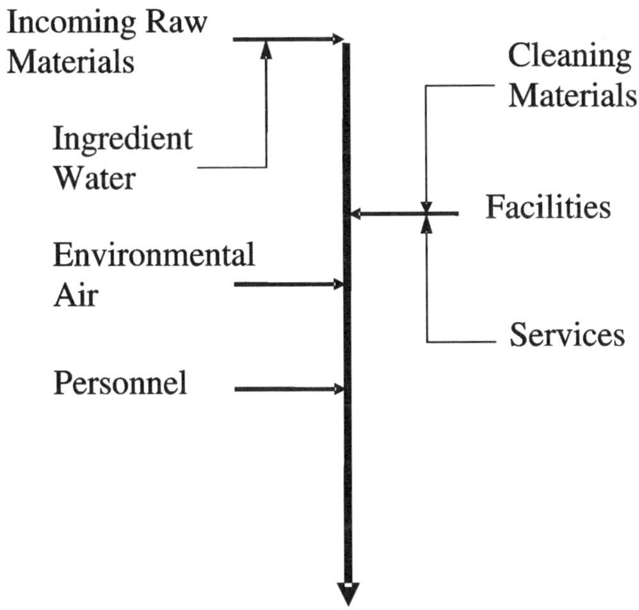

Figure 4.3. Generalized sources of microbiological contamination.

- Environmental air
- Personnel

Raw Materials

Raw materials are necessary and cannot be eliminated as sources of microbiological contamination. Pharmaceutical preparations should be formulated with raw materials that are unlikely to be sources of contamination. In principle, this means avoiding raw materials originating from plants or animal. Specifications for raw materials are generally defaulted to something in the order of 1000 cfu per gm or ml. It is often a difficult and contentious task to argue with vendors for the application of tighter microbiological specifications to raw materials used in critical applications.

Raw materials comprise the greater part of all topical formulation bases, thus, in terms of proportional contribution to microbiological contamination, they present the greatest threat.

Water is the base for lotions and creams, and so possibly the most significant source of contamination in any pharmaceutical preparation. It may be a source of contamination. It may be a vector for transmitting contamination. Its presence may encourage the proliferation of contaminants. Very significantly, water is the main habitat for Gram-negative types such as the pseudomonads, which are extremely metabolically versatile and potentially hazardous to topical preparations. *Stringent microbiological controls are applied to ingredient water used in pharmaceutical manufacture.*

The bases for gels are mainly cellulose derivatives, such as carboxymethyl-cellulose, prsenting only minor sources of microbiological contamination. For ointments, the base is generally something like petrolatum or white soft paraffin, very rarely sources or vectors of microbiological contamination. Although unlikely to be major sources of microbiological contaminants, it is essential they be sourced from suppliers who have given sound consideration to hygiene and microbiological control in the design of their manufacturing and distribution practices.

In addition to their bases, topical products also contain numerous excipients for emulsification, for maintaining suspensions, for absorbing water, for leaving protective films on the skin, and so forth. Some may be sources of microbiological contamination, for example, microcrystalline celullose, cetyl alcohol, stearyl alcohol, cetostearyl alcohol.

Water is the base and main raw material source of microbiological contamination to oral solutions and suspensions. In syrups the base may be sugar (sucrose), or in sugar-free syrups, sorbitol or hydroxypropoyl methyl cellulose. Sugar (sorbitol, hydroxypropyl methyl cellulose, etc.) may contribute to 60% or more by weight of the formulation. Sugar solutions up to about 65% by weight provide an excellent nutrient environment for molds, yeasts and other osmophilic microorganisms.

Microbiological specifications for sugar often contain special provision for limits on yeasts and molds, and may specify particular media for their detection. It is clearly impossible to apply the microbiological default specification of no more than 1000 cfu per g to sugar used at 60% concentration in a syrup with a finished product specification of (say) no more than 100 cfu per ml. A batch of sugar with contamination near the limit could be "passed" at incoming QC and then lead to "failure" of the finished pharmaceutical preparation after all other value has been added.

Tighter specifications must be applied, though it is often difficult to persuade suppliers to guarantee conformance to specification, particularly when pharmaceutical applications are only a small part of that supplier's output. Sometimes the willingness of a supplier to meet these demands is quite simply a cost function.

Emulsifying agents are essential to maintaining the stability of oral suspensions. They may be:

- Of natural origin (gelatin, casein, acacia, tragacanth, pectin, etc.)
- Finely divided solids (bentonite, aluminium hydroxide, magnesium trisilicate, etc.)
- Synthetic (sodium lauryl sulfate, benzalkonium chloride, etc.).

Those of natural origin are almost always microbiologically contaminated. Those of vegetable origin (such as acacia and tragacanth) often have large associated numbers of desiccation-resistant bacteria such as *Bacillus*. Other emulsifying agents rarely pose a microbiological problem.

Aqueous-based inhalations, may, according to the solubility of the active, be solutions or suspensions. All concerns regarding water in relation to other aqueous preparations apply equally to inhalations.

Microcrystalline cellulose is one of the agents most commonly used for maintaining insoluble actives in suspension. While microcrystalline cellulose is widely used in other pharmaceutical applications, tighter microbiological specifications may be required when it is used for inhalation preparations. This route of administration is thought to present a greater challenge to patient health than oral or topical routes. Thus the potential for particular microorganisms to be considered objectionable is greater.

Facilities

Facilities are rarely intrinsic sources of microbiological contamination. Poorly designed facilities, poor materials of construction and poor operational practices can, however, have a significant influence on levels of contamination originating from other sources associated with the facility.

New facilities for manufacture of pharmaceutical preparations can be designed to minimize internal sources of microbiological contamination and to facilitate hygiene. Numerous other matters have to be considered in new-facility design. Therefore, prioritization is always necessary for the inevitable compromises to have pragmatic and sensible outputs. Existing facilities may present vulnerabilities to microbiological contamination not been previously addressed through design. Microbiological control of such facilities is often difficult to resolve.

It is unfortunate — but true — that facilities that are difficult to clean cannot be cleaned properly. Microorganisms will survive and most likely proliferate in any area where dirt is allowed to gather. Visibility is one major focus point for cleanliness in facility design; it is difficult to reconcile hygiene with visible dirt, so walls, floors and ceilings should have light-coloured, smooth, cleanable, finishes. Floor-to-wall junctions should be coved, piping should be boxed in, cupboards and storage areas should be kept to a minimum and drains should be controlled.

Facility-related microorganisms may originate from cleaning materials (water), drains (foul water), services (cooling water, compressed gas). Most of these are water-related, again reflecting the importance of water as a source of contamination.

The cleaning process is intended to improve the cleanliness of the treated object. Sometimes it is ineffective. Many cleaning processes in pharmaceutical manufacture have been validated to avoid cross-contamination, with a focus therefore, on chemical cleanliness. It is quite possible that microbiological cleanliness may not be achieved on an object that is satisfactorily chemically clean. It is also possible that equipment and materials (e.g., water) used to obtain chemical cleanliness could leave the object in a poorer state of microbiological cleanliness than it initially was. This is not to say that there should be any greater stress on microbiological cleaning validation, perhaps only that there should be more emphasis on the microbiological validation of cleaning equipment and materials.

The main source of microbiological contaminants from cleaning is water. Water is typically the cleaning agent of choice. All cleaning water in facilities for the manufacture of pharmaceutical preparations must be of good microbiological quality. The degree to which cleaning water must be microbiologically controlled is a function of where it is to be used, what products and equipment it is being used in association with, and of the volumes to be used.

Potable water (generally to a microbiological specification of no more than 500 cfu per ml and absence of *Enterobacteriaceae*) is generally good enough for cleaning walls and floors in nonsterile manufacturing facilities. In some areas the municipal supply may easily meet these standards, but in others it is common to rechlorinate municipal or well supplies to ensure consistency, and to keep the distribution system in good order. Cleaning agents and disinfectants for walls and floors should be chosen carefully, even for these purposes. On occasions, personnel may become confused as to what is a cleaning agent, a disinfectant, and what is a proprietary cleaning agent–disinfectant combination. The author has experienced a facility in which product contamination with pseudomonads was traced to daily floor mopping with a cleaning agent thought to be a disinfectant but turned out to be something else.

Washing of product contact equipment and any product contact packaging components requires more attention to microbiological control than walls and floors. Solution residues may be easy to remove. Residues from suspensions (lotions and oral emulsions), syrups and semisolids (creams, ointments, gels) are difficult to remove. In the first instance the product residues must be removed to meet criteria for gross chemical cleanliness, along with materials that encourage the growth of microorganisms (e.g., sugar). Hot potable water would be the water quality of first choice, but it may be necessary to include the use of surfactants or cleaning agents. Residues of these surfactants may not be left on equipment and therefore the final rinses must be with water complying with the requirements of purified water (*USP* or *PhEur*). The microbiological limit applying to purified water is normally in the region of no more than 100 cfu per ml.

The most heavily contaminated water in pharmaceutical manufacturing facilities is in drains, the main habitat for Gram-negative microorganisms. These contaminants would be transmitted to other areas if there were a backflow, and on the feet and garments of operators who work close to drains (e.g., wash-bay operators). Drain locations should be minimized and facility design should ensure that drains are able to cope with expected volumes of water. Unused drains should be capped.

Cooling water is often of extremely poor microbiological quality (chlorination is often avoided to reduce the water's corrosion potential). Great care must be taken to ensure that it does not contaminate product through pin-hole or larger leakage in heat exchangers or jackets. Sometimes water of poor microbiological quality is used to cool glands on stirrers. Again, great care must be taken to ensure that O-rings and other seals are intact, in place and correctly specified. The author has experience of syrups rejected as a result of a pseudomonad contamination arising from a damaged seal in a water-cooled gland on a base-mounted stirrer in a manufacturing vessel.

Compressed gases are potentially potent sources of microbiological contamination in sterile facilities. This is because gas leakage, unlike water leakage, is not immediately visible. Gases are less likely to be major sources of contamination in nonsterile manufacturing facilities; nonetheless they may possibly lead to contamination, usually from Gram-positive desiccation-resistant species.

Environmental Air

Environmental air is an unavoidable source and vector for microbiological contamination. It is not a medium for proliferation, and many microorganisms find it inimical. Most airborne contaminants are desiccation resistant, Gram-positive *bacilli* and *cocci* carried and protected on fragments of dust, saliva proteins, etc.

Control of airborne contaminants is often one of the greatest perceived concerns in new-facility design. For most nonsterile pharmaceuticals, however, the risk of significant problems arising from airborne microorganisms is insignificant when compared to the risks from water.

Filtration, pressure differentials and air flows are used to control the contamination potential from environmental air. The degree to which these are necessary is a function of risk to product.

Personnel

Personnel are not only a source of microbiological contamination, they are also a vector for contaminants. People are mobile and unpredictable. They have their own microflora that can vary from person to person and occasion to occasion.

Even the healthiest, most hygienic person carries significant microflora — mainly *staphylococci*, *micrococci* and *coryneform* bacteria. Street clothing may carry a distinctly different microflora. Generally these microorganisms are harmless to the person and often to pharmaceutical preparations as well. The principles of containment apply to both the person (protective garments) and the process (line covers, laminar flow protection, etc.).

Illness changes the types of microorganisms originating from personnel, and increases their numbers. Pathogens like *Staphylococcus aureus*, *Streptococcus* spp., *Enterobacteriaceae*, etc. may be disseminated from open wounds, coughing, sneezing, skin-flaking, diarrhoea, etc.

4 CONTROL OF CONTAMINATION IN PRACTICE

Microbiological contamination control is not a precise science. Effective control usually requires an array of mechanisms — some are independent, some interact, some overlap, and some appear inconsistent. It is possible to argue logically against the standalone value of almost any and every single microbiological control measure, because it is only through their synergy that they are effective.

4.1 Facility Design and Mode of Operation

Successful elimination and minimization of sources and vectors of contamination rests with their identification and, largely, with facility design. The identification of sources and vectors has been addressed, but how are they to be eliminated or minimized? The answer to this question mainly lies in GMPs for the control of materials flow, people flow, air flow, and water.

Manufacturing facilities should be designed to ensure that materials are handled in a manner that affords control of microbiological- and cross-contamination. In the author's experience, cross-contamination is mainly a problem of powders, dusts, and solids. With liquids, ointments and semisolids, microbiological contamination is a far greater source of genuine problems.

Everything that comes into a facility will bring microorganisms with it. The first principle of control is to restrict the access points, preferably to one location for all incoming materials, and confine the contaminants as far as possible to this location. Much of what is effective in confining contaminants to incoming warehouses is achieved by negotiation with suppliers, with regard to delivery.

For instance, wooden pallets should not proceed beyond warehousing because they are potent sources of contamination. Cardboard boxes should be wrapped or shrouded in plastic to minimize the contamination carried on them. It is almost inevitable that cardboard boxes will be moved into the facility beyond the incoming warehouse: their contents should be supplied in inner plastic wrappers, allowing the cardboard to be discarded well away from any areas in which product may be exposed. A microbiologically well-designed facility has simple and clearly designated routes of movement of materials from warehousing to production — from minimally to better-controlled areas, with "dirty" wrappings shed at designated locations along the way.

Contamination from people is carried on shoes, clothing, hair and skin. The access of personnel to a facility for manufacturing pharmaceutical preparations must be controlled. There are innumerable variations on how personnel access control to facilities may be achieved, and to what extent control is necessary. The most stringent level is to have *all* personnel change out of their street clothes into a company uniform with dedicated footwear on entry to the facility. Thereafter, it may be necessary to change clothes again or to put on overalls for entry to designated manufacturing areas or other areas where product is exposed.

Secondary changing would be expected before entry into areas for the manufacture of liquids, ointments and semisolids; but probably not for filling areas (provided the filling machines have some localized protection). Facilities must be designed to accommodate changing rooms appropriate to the need for product protection.

Personnel working in the manufacture of liquids, ointments and semisolids should wear long-sleeved overalls, head covers, and cover up excessive facial hair (such as beards, moustaches or sideburns). Ideally they should wear gloves when they are handling the preparation, its raw materials, and equipment that comes into contact with the product.

Air is a significant potential source and vector of microbiological contamination to liquids, ointments and semisolids; particularly to inhalations where high microbiological standards are of greatest importance. This is due to the difficulty in treating Gram-negative infections of the lungs with antibiotics. The potential of air as a contaminant must be controlled by:

- Filtration
- Dilution through recirculation
- Positive pressure differentials
- Intact walls, closed doors and air locks

The air supply to manufacturing areas for liquids, ointments and semisolids should equally be filtered. The rating of filters needed to control contamination from air supplies to nonsterile manufacturing areas is not defined in the codes of GMP. Many manufacturers choose HEPA filtration, although HEPA filters are primarily intended for controlling the quality of air to sterile manufacture, and in sensitive applications in the electronics industry.

The necessity for HEPA filtration of air to topicals and oral liquids manufacturing applications depends on the quality of incoming environmental air and the prefiltration deployed. While HEPA filters are the only type with the significant retention of 0.3-μm particles (the approximate size of individual microorganisms), most airborne microorganisms are in fact carried in clusters on far bigger dust particles (about >5 μm in size) and will therefore be retained by less strictly rated filters. It is very unusual to find inhalations manufacturing areas for which the air supply is not passed through HEPA filters.

It is rarely economic, unless other factors such as cross-contamination come into play, to find single-pass air-filtration systems. It is more usual for filtered air to be recirculated through the filters, thus imposing a less stringent burden on the filters and diluting the challenge. Up to 80% of air is recirculated, sometimes more.

The rate of supply of air to manufacturing areas should provide airflow in an *outward* direction from the area requiring protection from airborne contamination; microorganisms are not equipped to move upstream against an airflow. Outward airflows are normally monitored through positive pressure differentials.

Recommendations in codes of GMP, such as 15 Pascals for sterile-area differentials, typically originate from experience past success rather than from any exact science. Ten Pascals is probably adequate for most nonsterile applications.

The final area of controlling contamination from air is the fabric and design of the facility. Air can be lost through walls if they are pervious, so impervious finishes are best used. If doors are opened, all positive pressure may be lost and contaminated air may enter an area that should be protected. Doors should be self-closing; and if the protection from airborne contamination is deemed important enough, they should be protected by air locks.

Water is probably the most significant combined source and vector for microbiological contaminants. Its control should be included in the design and operation of all facilities.

All water entering manufacturing facilities must be of potable quality, or must be treated (e.g., by chlorination) to bring it to these standards. This quality of water may be used for many applications, from drinking to equipment cleaning. If incoming water from a municipal supply fails to meet the customary microbiological standard of not more than 500 cfu per ml, little can be done to improve its quality except re-treatment in the pharmaceutical facility.

Ingredient water for pharmaceutical purposes must always involve some treatment of source water, to bring it to the pharmacopoeial standards for either purified water or for water for injection. Purified water is required for liquids, ointments and semisolids.

Typical treatments for preparation of purified water (e.g., deionization, reverse osmosis) improve the chemical quality of the water, but may not necessarily improve its microbiological quality. In certain circumstances, such treatments may actually lead to *poorer* microbiological quality.

Where distillation is used for the preparation of water for injection, the high temperatures involved give water for injection an intrinsically high microbiological quality.

Feed water pretreated with chlorination is used in processes for the preparation of purified water. Unfortunately, the presence of high concentrations of chlorine ions can *impair* chemical purification processes; excess chlorine is usually removed by passage of the rechlorinated water through carbon filters. Carbon filters, unless kept in good condition by recirculation of water and periodic backwashing, can themselves become a source of contamination. This may then be carried into the purified water-distribution system.

The microbiological quality of the water in systems for storage and distribution of purified water is maintained in a variety of ways. The water must always be kept in constant recirculation to prevent formation of biofilms on the inner surfaces of tanks and pipework. Valves and take-off points must be of the sanitary type. If the water remains stagnant there is an opportunity for microorganisms to multiply. The complete system must be periodically sanitised. This is usually best and most economically effected by using high temperatures, either at steam temperatures, or by recirculation above 80°C at intervals dictated by experience. To achieve these temperatures, heat exchangers should be included in the design of the distribution system.

Many purified water distribution systems include ultraviolet light stations to control microbiological quality. Care must be taken to ensure that these systems are working properly; it is quite possible that they only damage microorganisms, rather than killing them. Special media (e.g., R2A), low temperatures (e.g., 20–25°C), and extended times (e.g., 10–14 days) of incubation should be included in the validation of new systems to address this potential. The phenomenon of viable but non-culturable microorganisms after ultraviolet treatment and in general is well known.[2,5,6] Water-borne types are particularly difficult to culture; their mode of growth is not suited to the high concentrations of nutrients found in most general purposes media, nor are they suited to growing at temperatures above 30°C, and certainly not within a couple of days.

Some purified water distribution systems are operated at low (<15°C) or high (>40°C) temperatures to maintain high microbiological standards, but generally acceptable standards can be achieved more economically.

The other area of concern regarding water in facility design is drainage. Where water is left to stand, *Pseudomonas* spp. do not only survive but *proliferate*. Water should not be permitted to stand on equipment (particularly in crannies and crevices), on floors, or in sinks and wash-bays. Contamination spreads with water, forming films over surfaces and on the hands and clothing of personnel. Water-borne contaminants may be aerosolized by vibrations or when water falls more than a few centimetres. To restrict the opportunity for contamination from water, there should be air breaks of about 5 cm installed between equipment drains and tun dishes leading to foul drains.

4.2 Process Design and Mode of Operation

Many process-related aspects of manufacture of pharmaceutical preparations contribute to contamination, or conversely, to contamination control. These are generally product-specific, but some have intrinsic similarities amongst groups of products.

Figure 4.4 shows a process flow diagram for creams manufacture, given as an example for creams, lotions, gels and ointments. The major items of equipment are the fats vessel, the manufacturing vessel and the holding vessel. These three vessel types are general to all topicals manufacturing processes. Creams and lotions are at greater risk from microbiological contamination and proliferation than ointments and gels for formulation-related reasons.

Closed fats and manufacturing vessels are best from a microbiological standpoint because they are amenable to application of Clean-in-Place (CIP) systems. These usually afford opportunities for higher temperature cleaning than do manual systems.

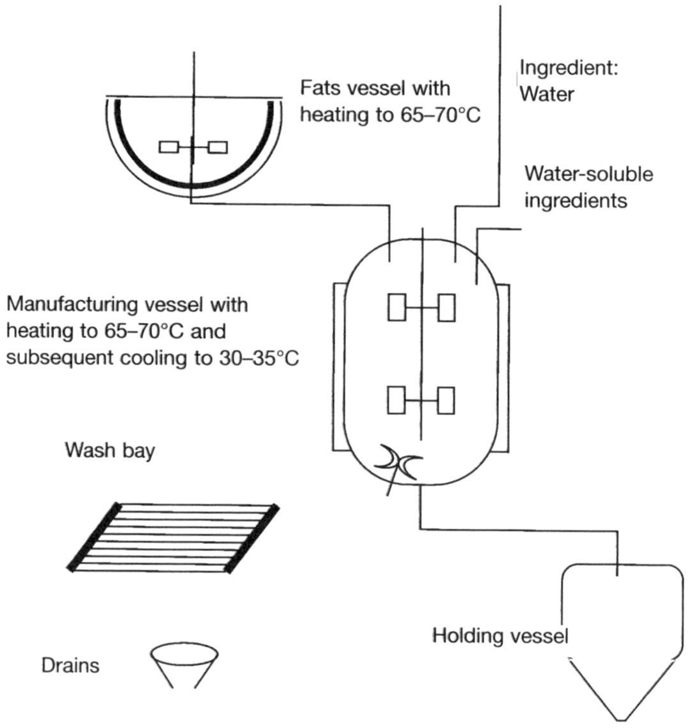

Figure 4.4. Generalized process flow for cream manufacture.

They are now commonly used, although the reason for this may have as much to do with facilitating the process, as to preventing microbiological contamination. Where open vessels are used and have to be cleaned out by hosing, there are significant risks of spreading and aerosolizing contamination throughout the area.

Materials are heated in both the fats and the manufacturing vessels to temperatures likely to inactivate most Gram-negative microorganisms, even if initially present in large numbers. Gram-positive spore-forming bacteria generally withstand these temperatures. These heating processes are pretty effective at minimizing the numbers of microorganisms introduced with ingredient water, raw materials and environmental air.

The drainage from the fats and manufacturing vessels may contribute to general microbiological contamination of topicals manufacturing areas. Control of this is not solely associated with leaving air breaks between floor and vessel drains to avoid backflow of polluted water into the vessels. It is also connected with the potential for drains to be blocked with solidified fats. Large volumes of water, which only have to comply with the microbiological limits for potability, may have to be used for the first stages of cleaning these vessels. If drains are not adequately provided with fat traps to ensure that the waste water is taken away fast enough, they may block and overflow. This results in the consequent spread of contaminants on floors and via the feet of personnel working in the area.

Most microbiological contamination in topicals manufacture arises from drains and wash-bays. The most significant vectors are personnel. Whereever possible personnel who work in wash-bays and with water should be required to wear dedicated footwear in these areas and rubber aprons to prevent their working garments becoming wet and contaminated. There is a significant risk of these personnel carrying contaminants to creams, gels, lotions and ointments during their transfer to the holding vessels, when the product has been cooled to temperatures at which microorganisms are likely to survive. Even products such as ointments and gels that are intrinsically unlikely to support microbial growth, have been known to become contaminated in holding vessels. There is evidence (in the public domain) of pseudomonads grown on films of condensed water lying on the surface of ointments and creams in holding vessels. Covering product surfaces in holding vessels with plastic "skins" is one means of minimizing this possibility.

Figure 4.5 shows a process flow diagram for syrup manufacture, as an example for solutions, suspensions and syrup. Again there are three types of vessel — ancillary, manufacturing and holding. As with topicals manufacture, the heating stages in manufacture are quite effective at minimizing microbial contaminants coming from ingredient water and raw materials.

Again, as with topicals, process-related sources of contamination are mainly from potable water used for cleaning and foul water from drains. Personnel and water are the most potent vectors of contamination. When possible, CIP systems provide better control of microorganisms than manual cleaning. It is an interesting and curious anomaly that vessels in which oral suspensions have been manufactured are most difficult to clean from a chemical *cross-contamination standpoint*; but vessels and pipework in which syrups have been manufactured are most difficult to clean from a *microbiological standpoint*. Validation criteria based on suspensions being the worst case (most difficult to clean), may not effectively address the risk of microbiological contamination after syrup manufacture.

If syrup residues are not cleaned from valves, pipework, filters, etc., they may create opportunities for environmental microorganisms to survive, proliferate, and create further problems. Screw-thread pipework connections must be avoided. There are frequently permanent hard-piped systems several metres long. They take oral liquids from manufacture to filling. Dead-legs and other foci for microbial proliferation in these pipework systems must be avoided. Such systems should preferably be designed as "demountable" for cleaning.

The manufacture of inhalation products requires more attention to microbiology than manufacture of topicals or oral liquids. This is because inhalation products couple a severe risk to patient health from the route of administration with a

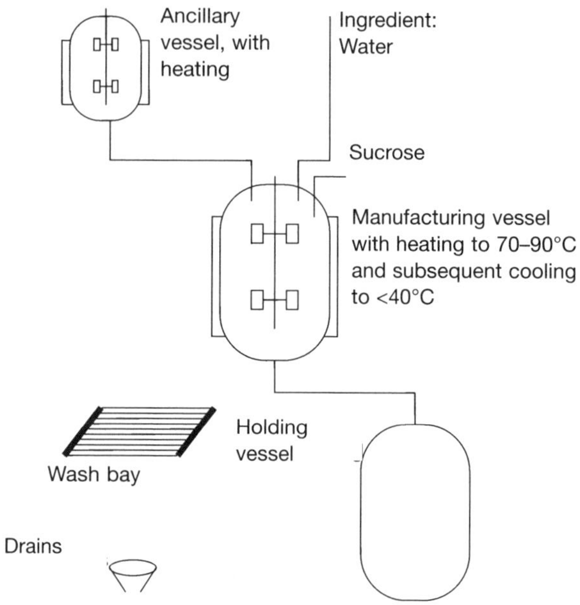

Figure 4.5. Generalized process flow diagram for syrup manufacture.

manufacturing process that does not necessarily incorporate any antimicrobial stages.

Manufacture of inhalations does not necessitate special manufacturing or filling equipment. It is possible, but quite unusual, for heating to be required for dissolution or suspension of the active ingredients. If manufacturing vessels are jacketed, it is to support high-temperature cleaning and sanitization. They are jacketed to support microbiological (rather than chemical or pharmacological) properties of the product.

Potable water is of insufficiently high microbiological quality for use in cleaning equipment during inhalations manufacture. Water complying with the microbiological limits for purified water must be used, even in the initial stages of cleaning product-contact parts and equipment. The concentration on cleanliness extends beyond the manufacturing process, to the control of the containers into which inhalations are filled, and to the applicators through which the products are administered to the patient.

4.3 Formulation-Related Microbiological Control

The problem of water content affecting the probability of microbial proliferation in pharmaceutical preparations has been discussed. Additionally, other formulation-related factors can influence microbial proliferation. Generally these are the inclusion of antimicrobial agents in formulations.

Gels and lotions frequently contain large concentrations of alcohols (e.g., isopropanol) for therapeutic reasons, such as rapidity of evaporation and provision of a "cooling" effect on the skin. More frequently, antimicrobial agents are included specifically for their preservative effects.

There is often some confusion as to exactly why preservatives are included in pharmaceutical preparations. They are included in multidose presentations to ensure that microorganisms — which inevitably contaminate preparations after they are first opened by the patient — do not proliferate to unacceptable levels before the patient finishes the course of treatment. The confusion arises because the same antimicrobial action providing "bathroom shelf" protection, can (a) prevent microorganisms multiplying over the unopened shelf life of the product, and (b) may also inactivate any microorganisms that contaminated the preparation in manufacture. However true this may be, and however effective the preservative system may be in particular preparations, it is totally unacceptable for preservation to be used as an excuse for poor microbial control in manufacture.

The measure of the effectiveness of preservative systems in pharmaceutical preparations is defined by the antimicrobial effectiveness tests described in detail in the pharmacopoeias.

The details of antimicrobial effectiveness tests differ between *USP* and *PhEur.* Nonetheless they follow very similar principles. A sample of product is inoculated with a specific culture. Subsamples are withdrawn at time zero and at intervals thereafter, and the number of viable microorganisms surviving is counted. According to the number of numerical logarithmic reductions at particular time intervals, the preparation passes or fails the test. This test must be done individually against a specified array of microorganims intended to represent the types of contaminant found in the bathroom-shelf environment. Some manufacturers support the inclusion of microorganisms from the local manufacturing environment in this array. It is unclear how these species can be considered relevant to the purpose intended for the inclusion of preservatives in formulations.

The advantage of performing a microbiological test of preservative activity over a chemical assay is that it takes account of any binding or inactivation of the preservative within the formulation that may impair its biological activity, but might not be detected by chemical assay. The antimicrobial effectiveness test must be done in new-product development and initially for determining product stability. Once the biological effectiveness has been established during new-product development, routine quality control may confine itself to chemical assay of preservative content.

In performing this test, it is preferable to inoculate product within its market container. It is mandatory that the volume that contains the inoculum is minimized in relation to the volume of product inoculated. This in itself may present serious technical difficulties. Good mixing is essential. This is often difficult to do in an ointment tube. Poor mixing may result in erratic results in the antimicrobial effectiveness test.

It should not be assumed that all aqueous-based nonsterile pharmaceutical preparations are preserved. There are pressures — regulatory and otherwise — to move to more nonpreserved nonsterile dosage forms. Although there are some containment systems allegedly designed to prevent ingress of microorganisms after delivering the dose, nonpreserved preparations are not generally suited to multidose presentations because they have no bathroom-shelf protection. They tend to be filled into single-dose presentations, such as nasal drops or sprays in plastic nebules. Soft gel capsules could be another route, but would require formulation in nonaqueous bases.

5 MICROBIOLOGICAL MONITORING OF THE MANUFACTURING FACILITY

The author does not know of any regulatory-defined microbiological standards for the manufacture of nonsterile liquids, ointments and semisolids. Standards for sterile manufacture are, on the other hand, well defined.

At first it might seem that the direction for defining microbiological standards for nonsterile manufacturing environments would be to "scale down" those already defined for sterile manufacture. However, this is not as easy as it would appear. Environmental microbiology standards for sterile manufacture are mainly quantitative (cfu per m^3, cfu per plate per hour, cfu per cm^2, etc.) and are limited to extremely low numbers, often zero. The extremely highly specified environmental controls applied in sterile manufacture make it possible to set and achieve these limits. When environmental controls are less stringent (as in nonsterile manufacture), the numbers of microorganisms not only increase, they also show more variability.

The consequences of the variability in numbers of microorganisms recovered from nonsterile manufacturing areas, and the relative weakness of the environmental control mechanisms, diminishes the effectiveness of quantitative standards. Limits are either set so high to accommodate the variability that they become meaningless, or set so strictly that they do not accommodate the intrinsic variability. They therefore result in frequent out-of-specification (OOS) conditions that cannot be sensibly rectified. In this latter case, the microbiological environmental monitoring programme most often falls into disrepute.

On the other hand, there are good arguments for setting limits for absence of the microorganisms, which indicate diminishing standards of hygiene, operator malpractice, system breakdown, etc. Gram-negative microorganisms are the greatest risk to liquids, ointments and semisolids. Although Gram-negative organisms are ubiquitous in nature in water and in drains, they are not all identically significant as potential contaminants of these preparations. Known Gram-negative pathogens should not be tolerated in manufacturing areas. However, there are other Gram-negative types, e.g., *Enterobacter agglomerans*, which need not always be of great concern.

The nonsterile environmental monitoring programme should balance the

(a) Types of Gram-negative organism (which may or may not be tolerable against the locations monitored)
(b) Type of product manufactured
(c) Severity of action required when they are isolated (see the scheme shown in Table 4.4)

Table 4.4 Suggested Approach to Monitoring Liquids, Ointments and Semisolids Manufacturing Areas for Gram-Negative Microorganisms

Location	Actions Required in the Event of	
	Detection of *P. aeruginosa*, *E. coli, Salmonella* or other confirmed pathogen	Detection of other Gram-negative species
Protected hygienically controlled locations such as manufacturing equipment, filling machines, etc.	• Suspension of manufacture • Action on product (rejection, recall)	• Corrective/preventive actions • Suspension of manufacture if persistent
Other less critical areas where sources of microbiological contamination might be anticipated, e.g., areas where water is used	• Corrective/preventive actions • Suspension of manufacture if persistent	Restrictions are not recommended because some isolation of these species is likely to be expected

Known Gram-negative pathogens are customarily addressed via attempted isolation of the indicator organisms, *Pseudomonas aeruginosa*, *E. coli* and *Salmonella*. Selective media are best used to ensure that these indicator organisms are not obscured by other, more resilient, microorganisms, which grow more rapidly and extensively on general media. The consequences of finding *Pseudomonas aeruginosa*, *E. coli* or *Salmonella* or other pathogens on manufacture or filling equipment are likely to be severe — batch rejection is a possibility, but this must be considered on a product-by-product, incident-by-incident basis. The consequences of finding them in a wash-bay, for instance, should not be as severe. Isolation of indicator organisms from these less critical areas should be seen as an early warning of a contamination source, which could lead eventually to manufacturing equipment or product contamination. Corrective and preventive actions against sources of contamination should be mandatory, and manufacture may have to be suspended if the problems persist.

Table 4.5 Recommended Frequencies for Nonsterile Environmental Monitoring

Type of Monitoring	Aqueous Inhalations	Creams and Lotions	Oral Liquids	Ointments and Gels
Microbial monitoring of surfaces by swabbing, or by contact plates	On each day of manufacture or filling	On each day of manufacture or filling	Weekly	Every two weeks
Viable microorganisms in air recovered by active sampling	On each day of manufacture or filling	Weekly	Every two weeks	Monthly

All Gram-negative microorganisms should be restricted from manufacturing equipment, filling machines, etc. General media are best used for this purpose, alongside the selective media used for the indicator organisms. The consequences of finding nonpathogenic Gram-negative species on manufacturing and filling equipment merit corrective and preventive action against the sources of contamination, but not necessarily action against product. As they may be indicative of the presence of pathogenic species at levels below the sensitivity of the detection methods used, manufacture may have to be suspended. It is to be expected that some Gram-negative microorganisms will be isolated from less critical areas, particularly those in which water is present. Unless isolates are of the indicator types or other confirmed pathogens, their importance should not be exaggerated.

Gram-positive types may be similarly dealt with, concentrating on isolation of *Staphylococcus aureus* on suitable selective media. Some incidence of ubiquitous environmental Gram-positive types (such as *Bacillus* spp. and *Micrococcus* spp.) is inevitable, and unless excessive, should not give too much cause for concern. Gram-positive types should be considered of greatest significance in locations from which personnel are excluded and locations (e.g., within filling cabinets) where personnel are expected to disinfect after rare and unusual intrusions. Isolation of *Staphylococcus aureus* in such locations, for example, means that *something is not happening in the manner intended.*

Table 4.5 indicates suitable frequencies and methods for microbiological environmental monitoring of facilities for manufacture of liquids, ointments and semisolids.

Actions and sanctions should be expected and applied when there are infringments to microbiological limits applying to the manufacturing environment for liquids, ointments and semisolids. The most likely actions arising out from out-of-specification or atypical nonsterile environmental results are those relating to the control of the process, or relating to operators and facilities.

The most frequently required actions are disinfection of an area or piece of equipment, or counselling to retrain personnel. These are typical corrective actions. *Corrective actions are defined in terms of fixing the immediate problem.* Additionally, attention must be given to preventive actions, for instance, replacing a defective item of equipment. *Preventive actions are defined in terms of making sure the problem cannot arise again.*

The most serious actions that can arise from an out-of specification result from microbiological environmental monitoring of nonsterile manufacturing areas are for product withdrawal (recall) or rejection. Neither action is likely to be required as a result of environmental data with no evidence of actual product contamination. Nonetheless it would be difficult to justify continuing manufacture of, for example, a paediatric syrup in equipment from which *E. coli* or *Salmonella* is repeatedly isolated, or to manufacture an inhalation in equipment from which *Pseudomonas cepacia* is isolated. Suspension of manufacture is a viable and probable option pending thorough investigation, diagnosis of the root cause of the problem and implementation of adequate corrective and preventive action.

REFERENCES

1. Ayerst, G. Water Activity — Its Measurement and Significance in Biology. *Int. Biodeter. Bull.*, 1: 13–26, 1965.
2. Byrd, J.J., Xu, H.-S., Colwell, R.R. Viable but nonculturable bacteria in drinking water. *Applied and Environmental Microbiology* 57, 875–878, 1991.
3. Casey, W.M., Muth, H., Kirby, J., Allen, P. Use of Nonselective Preenrichment Media for the Recovery of Enteric Bacteria from Pharmaceutical Products. *Pharmaceutical Technology*, 22, 1998.
4. Enigl, D.C., Sorrels, K.M. Water Activity and Self-Preserving Formulas. In *Preservative-Free and Self-Preserving Cosmetics and Drugs: Principles and Practice* (eds. J.J. Kabara, D.S. Orth). Marcel Dekker, New York, 1997.
5. Jones, D.M., Sutcliffe, E.M., Curry, A. Recovery of viable but nonculturable *Campylobacter jejuni. Journal of General Microbiology,* 137, 2477–2482. 1991.
6. McKay, A.M. Viable but nonculturable forms of potentially pathogenic bacteria in water. *Letters in Applied Microbiology* 14, 129–135, 1992.
7. Murray, P. *Manual of Clinical Microbiology*. American Society of Microbiology, Bethesda, MD, 1999.
8. Russell, M. Microbiological Control of Raw Materials. In *Microbial Quality Assurance in Pharmaceuticals, Cosmetics and Toiletries* (eds. S.F. Bloomfield, R. Baird, R.E. Leak, R. Leech). Taylor & Francis, London, 1988.

Chapter 5

Bioburden Determinations

Norman Hodges

CONTENTS

1 DEFINITIONS AND SCOPE

The word *bioburden* has been ascribed subtly different meanings, not only in scientific texts, but also in official documents. Many authorities use the word simply in a quantitative sense, i.e., in a way suggesting merely a determination of numbers, with little or no specific mention of types of organisms present. *ISO 11134*[1], for example, defines bioburden as "Population of viable microorganisms on a raw material, component, a finished product and/or a package." More commonly, however, the word implies both quantitative and qualitative characterization; thus, the glossaries of terms used in FDA[2] and the European Community[3] guidance on good manufacturing practice (GMP) explain *bioburden* in an identical manner:

0-849-32300-2/04/$0.00+$1.50

"The level and type (e.g., objectionable or not) of microorganisms that can be present in raw materials, Active Pharmaceutical Ingredients (API) starting materials, intermediates or APIs. Bioburden should not be considered contamination unless the levels have been exceeded or defined objectionable organisms have been detected." It is in this latter sense that the word will be used in this chapter.

Characterization of a microbial population could be taken to mean determining the relative numbers of *all* the different species present, and this implies identification of organisms. Clearly, identification of *all* organisms comprising the bioburden is not normally practicable, although the regulatory expectation of identification of organisms that regularly appear in successive batches of material, and comprise a major fraction of the bioburden is quite manageable. Unfortunately, details of identification procedures are outside the scope of this chapter, but reviews of the rapid automated methods now widely employed in the industry appear in this publication[4] and elsewhere.[5]

Although bioburden determinations are commonly applied to solid and liquid raw materials, intermediates and finished manufactured medicines, they are also required during manufacture and immediately prior to sterilization of medical devices. Because such devices cannot usually be sampled by the procedures employed for medicines, bioburden determinations require surface-sampling techniques like swabbing and the use of contact (Rodac) plates that are more commonly employed in environmental monitoring.[6]

2 FACTORS INFLUENCING THE BIOBURDEN AND TESTING REQUIREMENTS

The bioburden of a product will be influenced by a variety of factors including, but not necessarily limited to, the following:

- Microbiological quality of raw materials and components (including containers and packaging)
- The manufacturing environment, i.e., organisms present in the atmosphere, on working surfaces, plant and equipment and on personnel
- The nature of the manufacturing process, which might promote microbial inactivation or, alternatively, support microbial proliferation, depending on exposure of product or components to the various temperatures, pH values, organic solvents, etc. employed

Of these factors, the quality of the raw materials is, for many nonsterile medicines, the one that has the most profound influence on the microbiological quality of the finished product. The high standards required for manufacturing premises, and the application of procedures designed to restrict or eliminate opportunities for microbial proliferation in water-containing materials during manufacture, now largely ensure that little additional contamination is introduced during the manufacturing process itself. While there is a lot of compendial emphasis on bioburden determinations on finished products, investment of effort at the start of the process to fully characterize the raw materials is likely to be very cost-effective for control of final product quality.

While the major compendia describe procedures for both quantitative and qualitative bioburden tests (Table 5.1), it should be emphasized that adoption of these procedures alone, without regard to the impact of the above-mentioned factors that influence the bioburden, may not be sufficient to ensure regulatory approval. The scope of testing and validation required should be determined not merely by the unquestioning application of compendial tests, but by a consideration of the potential impact of all aspects of the manufacturing process, and the intended use of the product on its desired microbiological quality. It is well-established, for example, that "natural" materials of animal, vegetable or mineral origin are likely to possess a higher microbial count than synthetic ones, and in many cases they are more likely to contain potentially pathogenic organisms. On this basis, there would be a regulatory expectation that any excipient of natural origin would be subjected to detection tests for relevant objectionable organisms, regardless of whether the material in question was the subject of a pharmacopoeial monograph.

Potential objectionable organisms in different nonsterile dosage forms have been tabulated in a recent paper in *Pharmacopeial Forum*[7] and are listed in Table 5.2. It should be noted that this table is based directly upon the original paper, and reasons are not clear for the omission of certain potential pathogens from some of the categories, e.g., *Pseudomonas aeruginosa* is absent in five product categories for which the less hazardous species *P. fluorescens* is listed.

On this same basis of adapting the testing protocol to the nature and use of the product, it may be necessary, for example, to enumerate the presterilization bioburden of spores prior to a terminal sterilization process and, depending upon the validation scheme for the sterilization process (bioburden or overkill approach), quantify their resistance parameters. Total viable count (TVC) procedures would also need to be modified for a product likely either to contain strict anaerobes, or, by virtue of a low redox potential, support anaerobic growth during use, since standard TVC procedures only enumerate strict aerobes and facultative anaerobes.

Table 5.1 identifies the major compendial, international standard and regulatory documents pertaining to bioburden determinations. The first two of these categories in particular give detailed accounts of the procedures to be adopted for enumeration and detection of specified organisms, and it is not the purpose of this chapter to reproduce this information. Rather, it is intended to identify and explain the issues that impact on the selection of methods, reliability of data and relevant validation.

Table 5.1 Official Methods, Standards and Guidelines on Bioburden Determinations

Compendial Methods	International Standards	Regulatory and Professional Association Documents
EP 2003: • 2.6.12 Microbiological examination of nonsterile products (total viable aerobic count) • 2.6.13 Microbiological examination of nonsterile products (test for specified microorganisms) • 5.1.4 Microbiological quality of pharmaceutical preparations *USP 26*: • <61> Microbial limit tests • <1111) Microbiological attributes of nonsterile pharmacopeial products • <1227> Validation of microbial recovery from pharmacopeial articles • <1231> Water for pharmaceutical purposes (microbiological considerations) • <2021> Microbial limit tests — nutritional supplements	• ISO 8199 (1988) Water quality — General guide to the enumeration of microorganisms by culture • ISO 6222 (1999) Water quality — Enumeration of culturable microorganisms — Colony count by inoculation in a nutrient agar culture medium • ISO/TR 13843 (2000) Water quality — Guidance on validation of microbiological methods • ISO 11737–1 (1995) Sterilization of medical devices — Microbiological methods — Part 1: Estimation of population of microorganisms on products	• European Commission (2002) The Rules Governing Medicinal Products in the EC Vol 4: Good Manufacturing Practice (reproduced in the *Rules and Guidance for Pharmaceutical Manufacturers and Distributors*, 2002, U.K. Medicines Control Agency) • U.S. FDA Center for Drug Evaluation and Research (2001). Guidance for Industry Q7A Good Manufacturing Practice Guidance for Active Pharmaceutical Ingredients • Parenteral Drug Association (PDA) (1990) Bioburden recovery validation. Technical Report 21 • ASTM (1991) Standard Practices for evaluating inactivators of antimicrobial agents used in disinfectant, sanitizer, antiseptic or preserved products. Document E–1054–91 • U.S. FDA *Bacteriological Analytical Manual* online (2001) AOAC International

Table 5.2 Potential Objectionable Microorganisms in Nonsterile Dosage Forms*

Organism	Dosage Form							
	Oral Solid	Oral Liquid	Topical	Vaginal	Rectal	Otic	Nasal	Inhalants
E. coli	X	X						
Salmonellae	X	X						
Aeromonas caviae	X	X			X			
Aeromonas hydrophilia	X	X			X			
Aeromonas sobria	X	X			X			
Plesiomonas shigelloides	X	X						
Shigella spp	X	X						
Vibrio cholerae	X	X						
V. para-haemolyticus	X	X						
Yersinia enterocolitica	X	X						
Y. pseudo-tuberculosis	X	X						
Burkholderia cepacia		X	X	X	X	X	X	X
Pseudomonas fluorescens		X	X	X	X	X	X	X
P. aeruginosa		X	X					
Serratia marcescens		X	X	X	X	X	X	X
Staphylococcus aureus			X					
Staphylococcus saprophyicus				X				
Candida albicans				X			X	X
Klebsiella spp.				X			X	X
Proteus spp.				X				
Enterococcus spp.				X				
Moraxella catarrhalis							X	X
Aspergillus fumigatus							X	X
A. flavus							X	X
Cryptococcus neoformans							X	X

*Adapted from Cundell (2002)[7].

3 SAMPLING

Limited information is provided on sampling by the pharmacopoeial chapters or sections relating specifically to microbiological testing. The *USP 26* simply states, in <61> *Microbial Limit Tests*, "Provide separate 10 ml or 10 g specimens for each of the tests called for in the individual monograph." The *EP* is rather more helpful; it specifies in Section 2.6.12 the requirement for a sampling plan, mentions some of the factors that might influence plan design, and provides an example of a plan applicable to products in which the bioburden might be not be distributed uniformly. Additionally, the *EP* indicates the acceptability (or need) to prepare composite samples by mixing the contents of several containers to provide sufficient bulk of material for testing. This practice also facilitates sampling from all parts of the batch. Both pharmacopoeias, however, say little about such aspects as the timing of sample collection, operator training, maximum transportation times, storage temperatures, and validation.

Bioburdens are not static, and there is the potential for microorganisms not only to grow in water-containing materials but also to die in anhydrous materials, due to nutrient deprivation, desiccation, etc. The potential for proliferation is well recognized, and TVCs are usually performed on samples taken at intervals during vulnerable stages of the manufacturing process. The possibility of the count, or a specific fraction of it, declining, is less frequently considered. Gram-negative bacteria in particular are relatively sensitive to desiccation, so they might be recovered from a dry, raw material at a low level after a period of storage yet have been present in higher numbers immediately after receipt. This may have implications for the endotoxin load in parenteral products.

When testing raw materials, bioburden samples should be taken by personnel adequately trained in aseptic techniques in order to avoid contamination of the samples or the bulk material from which the sample is withdrawn. Sterile containers, measuring vessels, etc., and the use of sterile gloves are required along with other precautions including facemasks, hair covering and gowns, depending upon the susceptibility of the material to environmental contamination. To minimize the risk of contamination from airborne organisms, sampling should not take place in areas of high personnel movement or air turbulence. Clearly, the interval between sampling and testing should be minimized, but if transportation time to the laboratory is significant, evidence must be obtained to confirm that the bioburden does not change during that interval. Maximum acceptable transport or storage temperatures and times must be validated. Standard operating procedures (SOPs) should be written by a microbiologist, and, if necessary, they should be prepared in collaboration with laboratory staff familiar with the practical problems that might be posed by sample product. SOPs need to be product-specific. It is easy for a generic SOP describing sampling procedures to fail to address the problems that might arise from the physical nature of the material, or from the container in which it is stored. Aseptic sampling might be problematic if, for example, the product is not a free-flowing solid or liquid, or it is in a container like a sack that is difficult to open without introducing contamination and, possibly, even more difficult to reseal.

Samples should, of course, be representative of the bulk material or the manufactured batch, so it is normal to remove material from different parts of the bulk or from the beginning, middle, and end of the batch. The pharmacopoeias specify the quantity to be taken in the absence of specific indications in individual monographs, but these quantities, typically 10 ml or 10 g, might be reduced if, for example, the batch was a small amount of expensive material. In this case it would be necessary again to demonstrate that the quantity taken was adequately representative of the whole. Samples of manufactured products should be taken after primary packaging to ensure that the bioburden contribution from that source is taken into account.

The subjects of sampling schemes and the statistics that support them can be complex and Kuwahara[8] has provided a detailed account written from a microbiological perspective. The *EP* describes a sampling scheme (2.6.12) whereby five individual samples from a batch are tested independently, and assigned to one of three classes based upon the TVC recorded: acceptable, marginal and defective. The batch passes if none of the five values exceeds the monograph limit by a factor of ten or more (no defectives) and not more than two samples are in the marginal category with a TVC between the prescribed limit and ten times the limit. Despite its inclusion in the pharmacopoeia, it is not commonly used, however, because, for the great majority of raw materials and finished samples, the recorded bioburden is low or absent. This means that there are no samples in the marginal category, and increasing the sample number by a factor of five is an unnecessary waste of time and consumables.

The requirement to avoid extraneous contamination of specimens is just as important during the laboratory investigation of bioburdens as it is during specimen collection, so the facilities required are generally those of a Hazard Category 2 containment laboratory, and the manipulations should be undertaken in a HEPA-filtered, horizontal laminar flow cabinet (or biological safety hood for potentially hazardous specimens). The differences between laminar flow cabinets and biological safety hoods are not readily appreciated by many newly recruited personnel, and these differences, together with a knowledge of when each type of cabinet should be used, are important components of a staff training programme. It is necessary to emphasize:

- That a horizontal laminar flow hood only protects the product from operator-derived contamination and affords no operator protection
- The circumstances when a sample might contain a pathogenic organism. This obviously depends upon a company's product portfolio, but samples associated with vaccine manufacture and mammalian viruses as possible contaminants of certain biotechnology products are two possible generic examples
- The circumstances when a sample might require a biological safety hood for reasons unrelated to microbiology and infection, e.g. when the sample is toxic by inhalation

4 PREPARATION OF SPECIMENS

It is relatively straightforward to conduct a TVC and perform tests to detect the common objectionable organisms of pharmaceutical importance if the sample is a pure culture of healthy organisms suspended in a simple aqueous solution. Unfortunately, the reality is that samples frequently pose problems, because the organisms to be enumerated or detected are present in low concentrations as part of a mixture of organisms, and they may be starved and slow growing, dormant in the form of spores that are slow or difficult to germinate, or sublethally damaged. Furthermore, the sample might pose solubility problems, contain suspended solid particles to which the organisms are adsorbed and from which they are difficult to remove, or it may possess intrinsic antimicrobial activity that reduces the growth rate and ease of detection of organisms of interest. It is important to recognise the problems that might be posed by particular dosage forms, and the need to ensure that the procedure adopted does not present an opportunity for the bioburden organisms to reproduce simply by virtue of the long time period required for sample preparation. Delayed-release or enteric-coated tablets, for example, may not dissolve quickly in peptone water, so it may be necessary to aseptically triturate them using a sterile pestle and mortar. Bioburden testing of medical devices can often represent the greatest challenge of all because of the problems posed by sheer size and inaccessibility of internal surfaces to liquid culture medium.

The *EP* and *USP* give directions on the preparation of specimens for analysis. Both pharmacopoeias consider water-soluble products, nonfatty insoluble materials and fatty products. In addition, the *EP* covers transdermal patches and the *USP* considers aerosols. Useful information on sample preparation is provided by Millar,[9] the Parenteral Drug Association,[10] and, for medical devices, by *ISO 11737*.[11]

Where possible, samples should be dissolved in the culture medium or diluting fluid recommended in the pharmacopoeias. If the sample is insoluble in water it must be suspended, or, for fatty products, emulsified using surfactants and heat (< 40°C). In the rare cases where emulsification is not feasible, use of nonaqueous solvents is a possibility, but very few have acceptably low toxicity. Isopropyl myristate, for example, is recommended for solubilizing or diluting water-insoluble materials for compendial sterility tests (though not currently mentioned for bioburden determinations). Thorough validation, particularly with respect to toxicity to microorganisms, would be necessary if a case were to be made for using such a solvent. Insoluble solids are also suspended in an aqueous medium, and neither pharmacopoeia clarifies that the intention is normally to keep the sample in uniform suspension throughout the dilution and plating process, rather than to remove microorganisms from the suspended particles and count or detect them in the supernatant after the solid has sedimented. The latter strategy may occasionally be necessary if the density of suspended material is so great that colonies growing in, or on, the culture medium could not clearly be distinguished, but in this case validation to confirm the extent of removal of organisms from solid particles would be essential. Viscosity of the solubilized, suspended or emulsified sample should be considered, both from the perspective of volumetric errors resulting from incomplete drainage from pipettes, and, in the case of samples enumerated by membrane filtration, the problems of low flow rates. Measuring by weight overcomes the first, and raising both the temperature (to a value not exceeding 40°C) and the transmembrane pressure should increase flow rates.

Antimicrobial activity may be exhibited by components of the formulation; these may be either the active (e.g., antibiotics) or, more commonly, preservatives. While preservatives usually effect some degree of microbial killing, they will not render the product sterile, so bioburden determinations are still required. To avoid the problem of antimicrobial ingredients carrying over into diluting fluids and Petri dishes and retarding the growth and detection of any organisms that survived their presence in the undiluted product, the ingredient in question must be removed or inactivated.

Medical devices can rarely be handled in the ways commonly used for medicines. They are often large, always insoluble and possess many surfaces to which liquids cannot readily gain access. It is necessary either to immerse the device totally in culture medium or diluting fluid or subject it to surface sampling. Total immersion is preferred and may necessitate a supply of large, strong, sterile and sealable bags that can be filled with medium. The device is dismantled as far as practicable and any valves or taps must be opened to facilitate liquid entry. Surface sampling using contact plates or swabs is less desirable and requires extensive validation. It is necessary to confirm that the area sampled is representative of the whole, and to quantify the efficiency of removal of attached microorganisms. Ultrasonics, shaking with or without glass beads and flushing are other techniques that may be employed to facilitate removal, and these are considered in detail elsewhere.[10,11]

5 QUANTITATIVE METHODS

5.1 Choice of Method

Having prepared the sample in a suitable form for a TVC, the choice of counting method is the next consideration. The methods described in the pharmacopoeias are listed in Table 5.3, which also identifies their relative merits. Other methods e.g., surface drop (Miles and Misra method) and semiautomated techniques (e.g., spiral plating) may be used, if validated, and are considered elsewhere.[9,10]

The *EP* and *USP* differ in the guidance they offer on choice of method. The *EP* directs that membrane filtration or a plating method (pour plate or surface spread) should be used, and that the most probable number (MPN) method (called the multiple tube method in the *USP*) should only be selected if there is no satisfactory alternative. By contrast the *USP* directs that pour plates should be used for all sufficiently soluble or translucent specimens, and the MPN method used otherwise. Surface spread techniques are not mentioned at all in the *USP*. Membrane filtration, despite its widespread use in the industry — and its acceptability to the FDA — is not a technique specifically recommended in the *USP* section entitled *Total aerobic microbial count*, although it is mentioned in the preparatory testing section as a technique that may be used to deal with inhibitory substances.

The principal criterion for selecting a method should be its suitability for the specimen in question. Considerations such as speed, ease of operation and cost are secondary. Suitability, in this context, means how well the method will deal with

Table 5.3 Relative Merits of Different TVC Methods

Method	Advantages	Disadvantages
Membrane filtration	• Detects lower concentrations than all other methods • Removes inhibitory (antimicrobial) components	• Less suitable for viscous samples, emulsions or high particulate loads • Filter integrity may be compromised by solubilizers • Relatively expensive consumables
Pour plate	• Technically easy • Detects lower concentrations than surface spread or MPN	• Possibility of thermosensitive organisms being killed • Possibility of strict aerobes producing small colonies that are overlooked • Unsuitable for samples with high particulate loads and emulsions
Surface spread plate	• Colonies of aerobes and facultative anaerobes are relatively large and easy to count • Best for emulsions, insoluble solids and fungi	• Requires surface drying of agar to soak up sample • Difficult to obtain uniform spread of colonies: some may be confluent
Most probable number		• Relatively inaccurate and imprecise • Recommended by *EP* as method of last resort

problems like elimination of antimicrobial activity in the specimen and the accuracy and precision of the result. It is well established that the methods available do not all give the same numerical result and that the precision of each varies, but it is not possible to quote relative values for accuracy and reproducibility simply because these will depend upon the organism used for testing and the skill and experience of the operator. Because the various methods have the potential to exhibit different detection limits and different degrees of accuracy and reproducibility, once a method is established for a product it cannot be substituted at will by another, unless validation data show equivalence.

Although not specified routinely in the current *USP*, membrane filtration was described in a proposed new chapter <61> *Microbial Enumeration Tests*, as the most accurate method for TVCs.[12] Membrane filtration is stated to be the preferred technique for sterility testing because it is the most effective means by which intrinsic antimicrobial activity may be removed; the same logic applies in TVC determinations. The sample is passed through a filter, and soluble antimicrobial agents should be physically separated from organisms retained on the membrane. It is possible for the antimicrobial chemical to be adsorbed onto the surface of the organisms, although the *EP* recommendation of passing three × 100 ml volumes of rinsing fluid through the membrane is intended to address this problem. All TVC determinations are best performed in a laminar flow hood. This is particularly important in membrane filtration, because the use of vacuum pumps means that surrounding air may be drawn through the membrane before or after the liquid sample, and the recorded bioburden may be artificially high due to the airborne microorganisms. Filtration will detect lower concentrations of microorganisms than plating or MPN methods, and since the usual recommendation is that TVCs should be based upon plates containing not less than 30 colonies and sample volumes for filtration are typically 100 ml, the lower detection limit is approximately 0.3 CFU/ml. It is, of course, possible to detect lower concentrations by use of larger sample volumes providing that the filter remains unblocked. Slow filtration rates or membrane blocking may render filtration an unsuitable method for samples that are viscous or contain a high concentration of solid materials, and samples containing nonaqueous solvents or surfactants (solubilizers) may alter the membrane structure or porosity.

If the sample is known not to possess antimicrobial activity, a pour plate or surface spread plate is likely to be preferred to membrane filtration, because plating methods are easier to conduct, and usually quicker and less expensive since there is no expenditure on filter manifolds, membranes and rinsing fluids. The choice between pour plates and surface spread plates is often a matter of personal preference and are equally suitable for many types of samples. Pour plates can accommodate larger sample volumes (usually 1 ml, but up to 5 ml provided dilution of the medium is shown not to influence recovery) so they will detect lower cell concentrations. Thirty colonies derived from a 5 ml sample correspond to a lower limit of 6 CFU/ml. This contrasts with the situation for surface spread plates where the maximum volume of liquid that can be absorbed is 0.5 ml (corresponding to 60 CFU/ml) although smaller volumes of 0.1 to 0.25 ml (detection limits of 300 to 120 CFU/ml) are more commonly used. Other disadvantages of the surface spread method are that the agar surface needs to be dried in order for the inoculum liquid to soak into the agar and so provide discrete colonies. Control of surface drying (also referred to as overdrying) is important. If the agar is dried excessively the microbial recovery might be low, but failure to ensure adequate drying may result in bacterial multiplication in the liquid on the agar surface, and the plate becoming uncountable due to confluent growth. Confluence may also result from nonuniform spreading of the liquid over the surface. Against this, the spread plate eliminates the possibility inherent in the pour plate method that an artificially low result may arise if the bioburden contains either thermosensitive organisms that are damaged by the hot agar or a high proportion of strict aerobes (some fungi, *Bacillus* and *Pseudomonas* spp., for example) that produce colonies which, at the bottom of the agar, are so small due to inadequate oxygen diffusion that they are overlooked. The surface spread method might also be better when the sample is an emulsion or it contains suspended solids making it difficult to visualize small colonies within the agar. One aspect of the pour plate technique and the associated risk of killing thermosensitive cells is the problem arising from the use of microwave ovens to remelt solidified agar. This practice can result in the liquid at the very centre of the bottle becoming appreciably hotter than that nearest to the glass walls of the vessel. A prudent precaution to eliminate this problem is to cool the agar for a sufficient period in a 45°C bath to ensure a uniform temperature throughout.

The MPN method has little to commend it other than as a technique of last resort. It is relatively imprecise and has poor sensitivity. The table in the *EP* from which results are calculated indicates that as few as 3 CFU/ml may be detected, but as Green and Randell[4] pointed out, this means the real result could be as high as 17. The corresponding table in the *USP* is curtailed at the lower detectable limit of 23 CFU/ml which, at 95% confidence, could really be a value as low as 7 or as high as 129. The *USP* indicates that the method should be considered for samples that, by their nature, make colony counting difficult using pour plates, but since the MPN result is determined by recording the number of tubes showing growth (turbidity) in a series receiving different volumes of inoculum, any sample that makes colony counting difficult will probably also necessitate subculturing of turbid MPN tubes.The method is not recommended in either pharmacopoeia as a means of enumerating surviving organisms in a preservative efficacy test. It has, however, been described as the basis for an automated preservative efficacy test that may be used for screening large numbers of candidate preservative formulations.[13]

The lower limit of detection for a counting method must be specified on regulatory submissions and it is important that bioburden values lower than the stated detection limit are not recorded. It is sometimes possible for the observation of a single colony on a single plate to be inadvertently reported in this way. These detection limits, however, are difficult, if not impossible, to reconcile with the *USP* recommendations that bioburdens should, ideally, be based upon plate counts in excess of 25 to 30 (the *USP* is inconsistent: 25 is the value stated in <1227> and 30 in <61>). A count of 30 colonies on a pour plate that received the standard inoculum of 1 ml would normally correspond to 300 CFU/ml or gram of sample (assuming the normal sample preparation of 10 g dissolved in 100-ml diluent). A count this high would exceed the compendial permitted levels for certain product categories anyway, and would be well above the specifications for many raw materials. A count of zero colonies would commonly be recorded for many samples, and bioburden levels sufficiently high to conform to the *USP* minimum plate count would normally be expected only for materials of "natural" origin and herbal products. It is worth noting that while the *EP* identifies 300 colonies per plate as the upper limit consistent with good evaluation, it specifies no lower limit. Replication is another issue. A count performed in accordance with a compendial method must be plated in duplicate, and while it is useful to extend this to triplicate plates in order to achieve greater reliability for certain products that may be expected to give relatively high counts, triplication is simply a waste of effort when the material consistently gives zero colonies.

Contact (Rodac) plates and swabs that are employed for surface sampling of large medical devices are covered in Chapter 2.[6] Their validation should confirm an ability to recover all, or a consistent known percentage of, organisms artificially inoculated and dried onto the surface in question. One point worth emphasizing is that when cotton swabs are used, the organisms removed from the sampled surface and attached to the cotton fibres are not necessarily all transferred to the suspending fluid in which the swab is placed; this problem might be avoided if alginate swabs are used, since these dissolve completely in the presence of 0.1% sodium hexametaphosphate.

5.2 Diluents, Media and Incubation Conditions

Samples are typically diluted 10-fold prior to counting, and the diluting fluid should be of such a pH and osmolarity that there is no possibility of viability loss before the sample reaches the Petri dish. It is therefore necessary to confirm this as part of the validation program. Water is unsuitable as a diluent, not only because its pH can be outside the range of 6 to 8 — normally considered acceptable, but also because some sensitive organisms can suffer osmotic shock and die rapidly. Phosphate-buffered saline (PBS), saline-peptone or fluid soyabean casein digest medium (tryptone soya broth), are among the commonly recommended pharmacopoeial diluents. Wetting agents may be added to any of them if necessary, as may chemical inactivators, to neutralize antimicrobial activity. Of these, PBS will not support any significant microbial growth, whereas diluents containing peptone or protein hydrolysates will. Thus it is important to avoid substantial time delays between preparation of the diluted specimen and the final filtration or plating; the *USP* puts a limit of one hour on this interval.

Soyabean casein digest agar (tryptone soya agar; TSA) is the recommended and most frequently used plating medium for bacteria. It will also support the growth of yeasts and molds so the possibility exists of conducting both counts on the same plate simply by incubating at 30–35°C for 48 hours for bacteria and then at 20–25° for five days thereafter. Again this strategy would have to be validated. The more common approach is to use Sabouraud-dextrose agar (SDA) with (*EP*), or without (*USP*), added antibacterial antibiotics for the yeast and mold count. The *USP* also suggests potato-dextrose agar for this purpose, and its lower pH (3.5) compared with SDA (5.6) makes it more effective for the exclusion of bacteria. Two of the antibiotics the *EP* recommends for addition to SDA, benzylpenicillin and tetracycline, are thermosensitive, and must be added as sterile solutions after autoclaving; the alternative — chloramphenicol — may be added before. The value of adding antibiotics is questionable anyway, simply because most samples do not contain sufficient bacteria for the presence of bacterial colonies on the plate to pose a problem when enumerating the yeast and mold colonies. The *EP* defines the total aerobic count as the sum of the bacterial and fungal counts, but a correction may be made for any organisms growing on both media. The use of antibacterial antibiotics in the fungal medium renders this possibility very unlikely, but it could be the best justification for the practice of using just one medium and incubating at two different temperatures. Other categories of organisms that may be quantified in the bioburden include coliforms (usually plated on MacConkey's agar) and anaerobes (on reinforced clostridial medium, or one of several alternatives).

6 QUALITATIVE DETERMINATIONS

These are designed to give a "yes" or "no" answer to the question of whether a specific objectionable organism is detectable in a given sample. The word "detectable" is used rather than "present" because the situation here is similar to that in sterility testing. The possibility always exists that the organism of interest was present, but the conditions were not ideal, and it was not detected, so the mere passing of such a test does not *guarantee* that the organism was absent. The validation programme should confirm that the testing procedure will detect the organism of interest when it is present at a specified level in the product, but the sensitivities required by the *EP* and *USP* differ markedly in this respect.

The subject organisms of qualitative tests are described as "objectionable" and they are all potential pathogens, although they may also be significant as indicators of product quality. Currently, the *EP* and *USP* both describe tests for the absence of *Staphylococcus aureus*, *Pseudomonas aeruginosa*, *Escherichia coli* and *Salmonellae* in selected raw materials and finished product categories, and in addition, the *EP* has tests for *Enterobacteriaceae* and *Clostridia*. Other organisms have been suggested as candidates for inclusion in this category[7,12] and the first of these two papers lists 20 such organisms relevant in eight different dosage forms (Table 5.2).

The exclusion tests are not applied indiscriminately to large numbers of materials that are the subject of pharmacopoeial monographs. Rather, they are invoked for raw materials in which the presence of the organism is a realistic (or historical) possibility, e.g., *E. coli* and *Salmonellae* in gelatin, and for product categories where they might represent a significant infection hazard, e.g., *S. aureus* and *P. aeruginosa* in topical products. Presence of the organism might also be indicative of quality since the most likely sources of *E. coli* and *S. aureus* are faecal contamination and manufacturing personnel respectively; the semiquantitative test for *Cl. perfringens* in the *EP* is explicitly stated as a quality criterion.

The principle of the methods used for all of the detection tests is similar: the sample is prepared as for a TVC, then a portion of it is incubated in a liquid-selective enrichment medium. Such media normally support the growth of the organism of interest and suppress the growth of others, so that the selected organism increases both in absolute numbers and in relative terms compared with the other constituents of the bioburden. This increase makes its detection more likely when the liquid medium is streaked onto a selective agar which, after incubation, is examined for the presence of characteristic colonies. An interesting and significant difference between the *EP* and *USP* protocols is noteworthy. The *USP* method directs, for all four objectionable organisms, that the first container of liquid medium containing the sample should be examined for the presence of growth after incubation, and the remainder of the process is only undertaken if growth (turbidity) is observed. The *EP*, on the other hand, does not provide this option, and the sample only passes the *EP* tests when the whole procedure is complete and no colonies observed. Thus it is not at all uncommon, when conducting an *EP* test, to transfer clear liquid from the primary to the secondary enrichment medium, and subsequently to streak multiple plates when there is already a strong indication that nothing will grow on them.

The detection of the target organism is dependent upon laboratory staff recognizing a typical colony when the only guide to appearance that they may have available is a description in a pharmacopoeia that can, at best, be described as vague, e.g., "well-developed colourless colonies" is the *EP* description of *salmonellae* growing upon deoxycholate citrate agar. It is an irony that as the quality of pharmaceutical products has improved, the pathogens in question very rarely arise in the bioburden. Increasing numbers of laboratory staff may be scrutinizing plates for organisms that they have never actually seen growing. This is a strong argument for ensuring that those same staff regularly conduct validation tests to confirm the nutritive properties of the medium, so that they have an opportunity to see the target organisms. This validation requirement that the test procedures should be capable of detecting culture collection strains of the objectionable organisms is also an ideal opportunity for a digital photographic record to be taken of the characteristic colonies on each of the media employed. Such a collection of photographs is outside the scope of this chapter, but appears elsewhere,[14] together with an account of the characteristics and selectivity of the *EP*- and *USP*-recommended media and photographs of other organisms that might be the subject of false–positive identifications. Following the recognition of a suspect colony on selective agar, the organism may be subjected to a confirmatory test, e.g., indole test for *E. coli*, coagulase for *S. aureus* and oxidase for *P. aeruginosa*, although direct identification of the species using a commercially available product like API test strips or Vitek is more common. The confirmatory tests described in the pharmacopoeias are not necessarily absolutely specific for the respective organisms, although this is not always apparent from the text. The statement in the EP, for example, that "confirmation (of *S. aureus*) can be effected by suitable biochemical tests such as the coagulase test" disregards the fact that at least three other species of *Staphylococcus* are coagulase positive.[15] The identity of suspect organisms is most convincingly confirmed using an appropriate commercial testing scheme in addition to the pharmacopoeial so-called confirmatory tests. However, this assumes that samples that are, in fact, contaminated, do actually give rise to suspect colonies on the selective agar media, and there are several potential problems that might prejudice this outcome. Recognition or adoption of the following points and strategies might reduce the chances of a false negative result in qualitative tests.

- The major suppliers all produce culture media possessing the same names as those recommended in the pharmacopoeial tests, but in some cases, e.g., MacConkey's agar, there are differences in the formulation that might influence selectivity.
- Correct preparation and storage of media are essential for reliable results. The selectivity of some media might change quickly during storage, e.g., Bismuth sulphite, a *USP* medium for salmonellae, is most selective when freshly prepared, and becomes less so as it becomes green over 3 to 4 days storage at 4°C. This means the storage intervals during which the media permit reliable detection of low concentrations of the test organisms need to be validated. This may be of particular relevance because the ability of the media to detect low concentrations of the target organism depends not only upon its absolute numbers, but also on the concentration of other organisms that might be present. Thus, it may be easier to detect salmonellae at, say, 100 CFU/ml in a pure culture than to detect 10 or 100 times this concentration in the presence of a heavy load of other bacteria. Reduced selectivity as a result of prolonged or incorrect storage might result in the obscuring of a few salmonellae in the sample.
- The incubation conditions described permit relatively wide variations in both temperature and time, e.g., 30–35°C for 24 to 48 hours, and in some cases the appearance of the growing culture can change substantially within these limits. The hydrogen sulphide production that is typical of salmonellae, for example, might be apparent as a black colouration or precipitate during early growth in several of the common *Salmonella* media, but disappears after 48 hours at 35°C.[14] Similarly, the typical pink colour of *E. coli* on MacConkey's agar may substantially diminish after the first day of incubation. Note that differences in the incubation temperatures recommended in the *EP* and *USP* represent another source of incompatibility between the two methods. The *USP* essentially uses only two incubation temperatures for both quantitative and qualitative testing: 30–35°C for bacteria and 20–25°C for yeasts and molds. In contrast, the *EP* tests additionally require incubators at 35–37°C, 41–43°C, 43–45°C and 45.5–46.5°C.
- It is useful to be aware of the degree of selectivity afforded both by the media and the confirmatory tests. Just as the confirmatory tests are not absolutely specific for their intended organisms, so, too, are the selective agars less than perfect. Thus, *Proteus* species might grow on cetrimide agar and Baird-Parker medium, and species of *Edwardsiella* and *Citrobacter* may exhibit the typical textbook appearance of salmonellae on XLD medium. There are many other examples!
- For most objectionable organisms of pharmaceutical interest there exist a few strains that do not conform to the standard description. Most strains of *P. aeruginosa* exhibit a green or blue pigment, but a few possess an orange or brown pigment and few are nonpigmented. Similarly, about 5% of *E. coli*

strains do not ferment lactose, and about 5% of salmonellae produce little or no hydrogen sulphide when they might be expected to do so.

- The simultaneous testing of known positive and negative control organisms is recommended by the *USP* for the coagulase confirmatory test for *S. aureus*. This principle may usefully be applied, where possible, to other confirmatory tests and selective agars.
- There are commercially available test kits for the detection of certain organisms, e.g., kits based upon the coagulase test for *S. aureus* and salmonellae; these are generally more convenient to use and more reliable than traditional textbook descriptions of the test procedures.
- Salmonellae, in particular, may pose problems of recognition, because certain harmless organisms may mimic salmonellae in both the selective agars and the triple sugar iron agar that is recommended as part of the confirmation process. Even API and Vitek results do not always afford a high degree of confidence in detection based upon biochemical tests. Agglutination tests using *Salmonella* antisera may provide a definitive answer.

7 VALIDATION

In addition to the minor aspects of validation, a bioburden validation programme has principally to demonstrate that the procedures routinely used are capable of:

- Accurately and reproducibly enumerating low concentrations of organisms contained in, or, in the case of devices, on, the surfaces of samples or products
- Detecting low levels of specific objectionable organisms in products
- Adequately neutralizing any antimicrobial activity associated with the product and that any chemical inactivator employed is not itself toxic

The principle of these validation procedures is simply that the sample is inoculated with a known number of challenge organisms, and subjected to the routine method for TVC or detection of objectionable organisms. The sample is considered to possess no antimicrobial activity and the process is validated if a minimum designated proportion of the inoculum is recovered, or, in the case of qualitative testing, the objectionable organism is detected. The detailed methods are covered in the *EP* (2.6.12 and 2.6.13), *Effectiveness of culture media and validity of counting method* and *Nutritive and selective properties of the media and validity of the test* respectively. The *USP* describes validation in <61>, *Preparatory Testing*, and in <1227>, *Validation of Microbial Recovery from Pharmacopeial Articles*.

Obtaining reliable TVC values from medical devices, complex formulations or sparingly soluble raw materials is much more difficult than demonstrating accuracy and reproducibility in counting simple aqueous suspensions of pure cultures. It is clearly desirable that an operator is capable of demonstrating the latter skill before attempting the former. What constitutes an adequate level of competence in debatable, but Sandle* has described a simple statistical analysis of pour plate data that can be used for technician qualification. An alternative "rule of thumb" is that replicate dilution and plating of a bacterial (but not necessarily a fungal) suspension, should yield a coefficient of variation of <10% in the resulting colony counts.

The number of colonies counted on a plate influences the accuracy of the recorded result, and the pharmacopoeias indicate that between 30 and 300 colonies on a standard 9-cm plate is appropriate for most bacteria and *Candida albicans*. Since this range is not optimal for all environmental isolates, however, or for many yeasts and molds or for counts on 47-mm diameter membrane filters, it is necessary to validate the countable range. A procedure for doing so is in *USP* <1227>.

Routine challenge organisms to be used for validation are described in the pharmacopoeias, although it would be appropriate to use additional or alternative organisms that

- Are regularly isolated during environmental monitoring
- Regularly constitute a significant fraction of the raw material or finished product bioburden
- Are critical to the process with which the bioburden sample is associated

Pure culture rather than mixed culture inocula are recommended by the *USP* for each of the organisms selected, but the *EP* method requires a mixed inoculum.

The compendial recommendations are inconsistent on the number or concentrations of organisms to be inoculated into the product or the percentage recovery values considered acceptable. The *EP* directs that suspensions should contain about 100 CFU/ml, while the PDA recommend both a low level (<100) and a high level (10^3–10^4) inoculum. The *USP* requires 1 ml of a 1000-fold dilution of an overnight culture to be added to the first dilution of the product (100 ml). The concentration of an overnight culture of many bacteria is approximately 10^9 CFU/ml, so the *USP* requirement corresponds to a final inoculum concentration of approximately 10^4 CFU/ml. These recommendations apply both for quantitative and "absence of" testing, so it is worth reemphasizing that in qualitative tests it is often much easier to detect an objectionable organism from an inoculum of 10^4 than from 10^2 organisms. Thus the *EP* validation is, in this respect, more rigorous than that of the *USP*. The FDA would expect a low inoculum in the region of 10 to 100 CFUs.

If there is no reason to suspect that the sample will exhibit any intrinsic antimicrobial activity, it is sufficient just to inoculate it with a known number or concentration of the selected challenge organisms and demonstrate that the

*http://www.pharmig.org.uk/pages/news/issue5/big.html.

minimum designated percentage is consistently recovered (usually in three batches). The percentage recovery required varies: the PDA and *USP* direct that a minimum of 70% is acceptable, the *EP* states that the count "should differ by not more than a factor of five" from the value derived from the inoculum, and the American Society for Testing and Materials recommends statistical analysis to identify significant differences.[16] The *EP* acceptance of 20% recovery is out of line with industry practice and regulatory expectations.

There are two strategies available for validating the recovery of microorganisms from medical devices, but both have their drawbacks. The first approach is that commonly used for raw materials and finished medicines whereby the device is inoculated with a known number of CFUs which are dried (under HEPA-filtered laminar flow conditions) onto the surface and subsequently recovered by one or more of the following: swabbing; immersion and rinsing; ultrasonics; and glass beads.[10,11] The problem with this is that many organisms, particularly Gram-negative bacteria, are susceptible to desiccation and may be killed by the drying process itself, so there is always doubt whether a recovery of less than 100% is due to inadequate removal or bacterial killing. This is unlikely to be such a problem with Gram-positive challenge organisms, especially sporeformers. The alternative method described in ISO 11737[11] is termed validation using repetitive recovery. Here, the product is not artificially inoculated, but its naturally occurring bioburden is enumerated by subjecting the product to repeated cycles of the recovery procedure until no more organisms are removed. This process is time-consuming and, as with a deliberate inoculation (spiking) method that gives a low recovery, there is uncertainty at the end about whether there are yet more organisms to be recovered. The ISO 11737 suggests coating the surface of the device with molten agar which, when set and incubated, should permit residual organisms to give visible colonies. Producing a coating of uniform thickness and incubating in conditions that prevent the coating drying complicate this approach.

For some products possessing antimicrobial activity membrane filtration may not be an option and dilution of the sample or the use of chemical inactivators may be necessary. The *EP* lists some common inactivators in 2.6.13, but more comprehensive lists appear elsewhere.[17] It is necessary both to demonstrate that the inactivator does effectively eliminate the antimicrobial activity, and that it is not toxic to microorganisms, so any validation process should therefore involve three viable counts. The first is for the inoculum suspension of the challenge organism, and the second and third are the organism in the presence of the inactivator with and without the product sample (testing inactivator effectiveness and toxicity respectively). The acceptance criterion is again not less than 70% of the control count recovered throughout a minimum 30-minute period of contact between the challenge organism, inactivator and sample. Because inactivator formulae frequently contain surfactants like lecithin and tween that disaggregate bacterial clumps, it is not uncommon for the recovery value to be >100%.

8 DOCUMENTATION AND DATA ANALYSIS

It is necessary both to document the procedures fully and to establish a system for recorded and trending the data, particularly quantitative bioburden determinations. The documentation for both routine methods of bioburden determinations and the associated validation processes should comprise standard operating procedures that record materials and equipment, methods and acceptance criteria, together with identification of personnel responsible for obtaining and approving the data. The account of the methods must include the details of all measurements and measuring equipment, incubation conditions, the composition and source of all diluents, inactivators and media, and the source and maintenance of reference organisms.

The manner and detail in which results are recorded demand careful consideration. The primary data, i.e., the individual plate counts, together with the calculation of means and the final bioburden value that is the product of the mean and appropriate dilution factors, should all be recorded. It is desirable that bioburden counts are recorded in a manner that then makes them amenable to charting, trend, and possibly statistical analysis. It is important that a numerical value is recorded if at all possible. If, for example, results were regularly recorded for water as <1 CFU/ml it would be difficult, if not impossible, to ascertain how, or whether, the quality was changing. Quantitative bioburden data may conveniently be recorded using statistical process control charts, the uses of which, in a pharmaceutical manufacturing context, are described by Ingram and Cochrane.[18]

The colonial morphologies of the major organisms that routinely constitute the bioburden are likely to be familiar to the personnel who regularly undertake the work, and it is useful to record a presumptive identification based upon visual recognition of colonies. This recognition should comprise part of the staff training program. Gram-negative organisms are often of particular interest or concern, either as potential pathogens, or as a source of endotoxins. Consideration should be given to the possibility of including in SOPs a statement that any presumptive Gram-negative isolates should be examined under Gram stain, and if still Gram-negative, identified to species level. Well-defined procedures also need to be in place describing the application of out-of-specification and atypical analytical results procedures and the manner in which results are transmitted for the purpose of batch release.

9 REFERENCES

1 ISO 11134. *Sterilization of Health Care Products — Requirements for Validation and Routine Control — Industrial Moist Heat Sterilization*, 1994.

2 FDA Center for Drug Evaluation and Research. *Guidance for industry Q7A good manufacturing practice guidance for active pharmaceutical ingredients*, 2001.

3 European Commission. *The Rules Governing Medicinal Products in the EC, Vol 4: Good Manufacturing Practice* (reproduced in the Rules and Guidance

for Pharmaceutical Manufacturers and Distributors 2002 U.K. Medicines Control Agency), 2002.
4 Green, S., Randell, C. Rapid microbiological methods explained. This volume, 2004.
5 Wassall, M. Rapid enumeration and identification methods. In *Industrial Pharmaceutical Microbiology: Standards and Controls* (eds. N. Hodges, G. Hanlon), pp. 5.1–5.33. Euromed Communications, Haslemere, U.K., 2003.
6 Halls, N. Microbiological environmental monitoring. This volume, 2004.
7 Cundell, A.M. Comparison of microbiological testing practices in clinical, food, water and pharmaceutical microbiology in relation to the microbiological attributes of nutritional and dietary supplements. *Pharmacopeial Forum*, 28, 964–985, 2002.
8 Kuwahara, S.S. Microbiological based statistical sampling. In M*icrobiology in Pharmaceutical Manufacturing* (ed. R. Prince) pp. 485–505. Parenteral Drug Association, Bethesda, MD and Davis Horwood International Publishing, Godalming, U.K., 2001.
9 Millar, R. Enumeration of microorganisms. In *Handbook of Microbiological Quality Control: Pharmaceuticals and Medical Devices* (eds R.M. Baird, N. Hodges, S. Denyer), pp. 54–68. Taylor & Francis, London, 2000.
10 Parenteral Drug Association. Bioburden recovery validation. Technical Report 21. *Journal of Parenteral Science & Technology*, 44, (6), 324–331, 1990.
11 ISO 11737. Sterilization of medical devices — Microbiological methods — Part 1: Estimation of population of microorganisms on products, 1995.
12 Pharmacopeial previews: <61> Microbial enumeration tests; <62> Microbiological procedures for absence of objectionable microorganisms; <1111> Microbiological attributes of nonsterile pharmacopeial articles. *Pharmacopeial Forum*. 25, 7761–7791, 1999.
13 Fels, P. An automated personal computer-enhanced assay for antimicrobial preservative efficacy testing by the most probable number technique using microtitre plates. *Pharmaceutical Industry*, 57, 585–590, 1995.
14 Hodges, N. Pharmacopoeial methods for the detection of specified organisms. In *Handbook of Microbiological Quality Control: Pharmaceuticals and Medical Devices* (eds. R. Baird, N. Hodges, S. Denyer), pp. 86–106. Taylor and Francis, London, 2000.
15 *Bergey's Manual of Determinative Bacteriology,* 9th ed. Williams & Wilkins, Baltimore and London, 1994.
16 American Society for Testing and Materials. Standard Practices for Evaluating Inactivators of Antimicrobial Agents Used in Disinfectant, Sanitizer, Antiseptic and Preserved Products, Document E 1054–91, 1991.
17 van Doorne, H. A basic primer on pharmaceutical microbiology. In *Microbiology in Pharmaceutical Manufacturing* (ed. R. Prince), pp. 71–123. Parenteral Drug Association, Bethesda, MD and Davis Horwood International Publishing, Godalming, U.K., 2000.

18 Ingram, M., Cochrane, T. Statistics and statistical process control in pharmaceutical microbiology. In *Industrial Pharmaceutical Microbiology: Standards and Controls* (eds. N. Hodges, G. Hanlon), pp. 5.1–5.33. Euromed Communications, Haslemere, U.K., 2003.

Chapter 6

Materials of Construction and Finishes for Safe Pharmaceutical Manufacturing

Dennis Fortune

CONTENTS

0-849-32300-2/04/$0.00+$1.50

1 INTRODUCTION

1.1 Scope

This chapter addresses the importance of an integrated design approach to microbiological contamination control in pharmaceutical manufacturing areas and clean-room construction. Other industries where clean-room manufacturing is highly regulated (food, beverage, electronics, banking, nuclear etc.) will all benefit from the information collated here.

We deal with the fit-out of the manufacturing enclosure and review the requirements for good architectural detailing and selection of appropriate construction materials and finishes.

The design of the layout for the clean room is briefly reviewed in terms of configuration, equipment layout, general operability and good manufacturing practice (GMP). Some typical examples are used to illustrate some of the key layout issues affecting clean-room design.

Detailed guidance is also given on the architectural design issues relative to the selection, performance and architectural detailing of construction materials and room finishes. In addition we look at the requirements for fixtures and fittings. Summary tables provide guidance on the standard of architectural detailing and surface finishes required to meet the appropriate clean room classifications.

The chapter deals with the interaction of cleaning and disinfection methods and materials on the fabric and finishes of the clean room. It also addresses the methods used for supplying the correct air quality and physical integration of this requirement with the clean room.

In architectural terms, the definition of a clean room is taken as the room enclosure, including any openings, penetrations, etc., to be constructed with sufficient integrity and detail so as to provide a microbiologically contained environment, which can be maintained within particulate and microbiological limits, in order to give the product, and sometimes the operator, protection from contamination.

2 BACKGROUND

Clean-room design has developed considerably over recent years, and in particular to meet the requirements for a high-quality contained environment, not only for the pharmaceutical industry, but also for the nuclear energy and electronics/microchip industries. The demand for good-quality clean rooms has led to the development of a multitude of specialist companies and products catering for the need for specialist methods of construction, and the exacting regulatory requirements for room finishes and components.

With the growth in increasingly biologically active drugs and the development of biotechnology, more onerous and exacting requirements are now applied by the various regulatory authorities. At the same time, room construction and finishes are becoming ever more sophisticated and expensive, making industry all too aware of the spiraling costs of clean rooms. Methods of combating rising costs have already been introduced. A good example is the series of *Baseline Guides* developed by the International Society For Pharmaceutical Engineering (ISPE) in cooperation with the U.S. Food and Drug Administration (FDA). These guides are aimed at setting the minimum standards for a fit-for-purpose design, which will still comply with regulatory criteria.

In addition, the introduction of barrier technology means that the standards for background clean rooms can be rationalized, due to the totally contained approach. Cost benefits gained must be balanced against the cost of the isolator equipment, which is expensive.

There is also another approach, which limits the size of the clean room: having an efficient layout, and minimizing the room area by removing all activities and equipment not essential to the process to an adjacent but lower-quality space. The smaller the volume of the high-quality finished and highly serviced room, the less expensive it becomes.

The clean room classifications used in this chapter are those of the International Organization for Standardization (ISO) and in particular ISO 14644–1.

3 CLEAN-ROOM LAYOUT

3.1 Layout Issues

The clean-room layout must accommodate the process requirements and equipment layout while maintaining good levels of access for operability, maintenance and personnel, material and component movements. It must also address suitable access for cleanability and disinfection.

The layout should prevent product cross-contamination, environmental microbiological contamination, and address the issue of contamination at any product or operator interface. In addition, it should be possible to easily remove

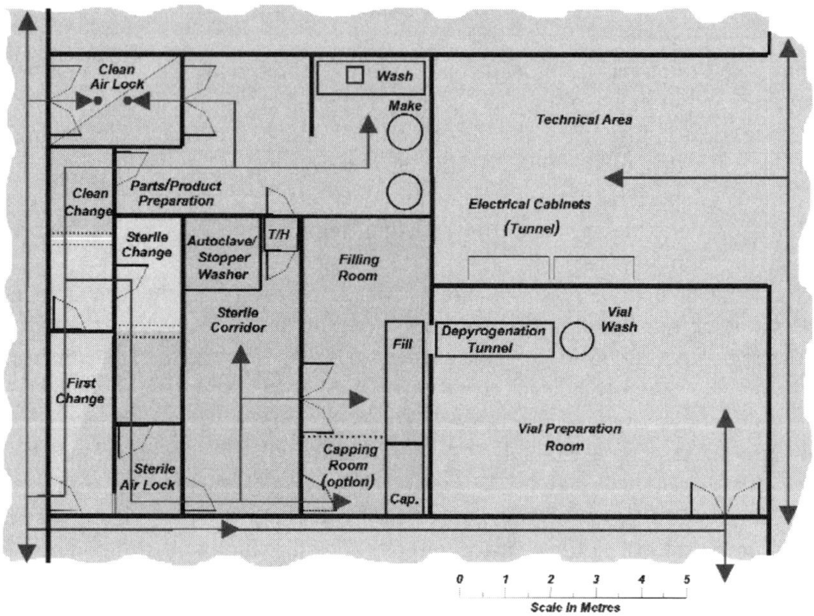

Figure 6.1. Typical clean-room layout personnel flows.

waste from the process or operation, without it again passing areas where the product is exposed.

Personnel flow through the facility is a key issue contributing to successful operation of the clean room. Personnel flow routes should be clearly defined, with smooth transitions between garmenting zones from facility entrance, offices and general plant through to operational areas. Product, material, equipment and personnel flows can usefully be illustrated on equipment layout drawings (Figure 6.1).

The architectural design detailing and finishes should provide a contained environment with the selected room finishes, which enhance hygiene, microbiological environmental control and safety levels. In addition, the design detail and finishes specifications must comply with the relevant fire codes and building regulations.

The increased demands for visual communication between clean-room areas must also be addressed, and the construction of the clean-room fabric should be able to accommodate flush glazed viewing panels.

Area clean-room classification, and the identification of other hazards should also be reviewed for their impact on the clean-room design. Chemically resistant or antistatic finishes may be required and, in particular instances, explosion relief panels may have to be utilized.

The provision of services into the clean room and the integration of floor and wall-mounted equipment should also be carefully considered and be able to provide neat, cleanable, sealed, interfaces with the room fabric and finishes.

4 CONSTRUCTION

4.1 Structure

The structural framework and building fabric selected for the building can have considerable impact on clean-room design and its performance. In particular, the intrusion of structural features — such as expansion joints and structural columns or beams projecting into clean areas — should be avoided, if at all possible.

Equipment interfaces with the building fabric and finishes should be minimized where the integrity of the room is critical. This is most easily achieved by locating all nonessential equipment outside the clean room.

Where it is critical for equipment to be located in the clean room, then an acceptable alternative would be to build the equipment into the clean-room wall to allow operator access and loading to the front of the machine. The rear of the machine could be located in a utility area with the majority of the utility and service connection pipework restricted to this area.

4.2 Construction Materials

Common materials used in the base construction of floors, walls and ceilings of clean rooms are listed here. The selection of materials will, however, depend on the type of facility in which the clean room is situated.

The more common types of facility are as follows, together with key requirements for the building fabric and finishes.

(a) Primary (Bulk) Facility

The large-scale manufacture of compounds and intermediate products usually requires the movement of large quantities of materials, which are sometimes heavy, with forklift and hand-pallet truck. Finishes must be robust, solidly constructed and hard-wearing. Generally the classification of the clean rooms will be of a lower quality.

(b) Secondary (Finishing) Facility

The final form of drug manufacturing generally involves the movement of smaller quantities of mostly lighter materials. Finishes tend to be lighter in construction, but they are generally of a higher quality and often required to have an element of flexibility. The clean-room classifications will be higher quality.

(c) Biotechnology Facility

The biotechnology facility can involve both the primary and secondary stages of manufacture, and therefore construction and finishes requirements can be from either (a) or (b).

(d) Research and Development Facilities

Small-scale and pilot plant-type operations usually involve much smaller quantities of material. They are closer to laboratory scale operations, which define the standard required of high quality and usually flexible requirements. Clean-room classification will therefore be of the higher quality.

4.3 Design Details

In detailing the construction of the clean-room floors, walls and ceilings, the following fundamental aspects must be clear in the designer's mind.

- The materials used for finishing the room surfaces must be nonshedding, non-porous and resistant to sustaining microbial growth.
- The finished surfaces must be hard, smooth and easy to clean with no ledges and minimal surface joints.
- The junctions of room surfaces must be carefully detailed to avoid inaccessible corners, preventing dust accumulation and facilitating cleaning.
- Coving at floor-to-wall, wall-to-wall and wall-to-ceiling junctions should be detailed and the radius in the range of 40 to 75 mm, depending on the materials chosen (see Figure 6.2).
- The selected finishes must be able to withstand repeated disinfection with the cleaning methods and disinfectants identified in the plant-cleaning philosophy.
- The integration of equipment and services into the room fabric must take account of all of the above requirements.
- The number of openings in the clean-room fabric should be minimised. Doors and vision panels must be detailed flush and form a continuous surface with the adjacent wall.
- Door hardware (furniture) should be minimized with the use of concealed door-closer mechanisms and flush push plates. Door hardware should have a

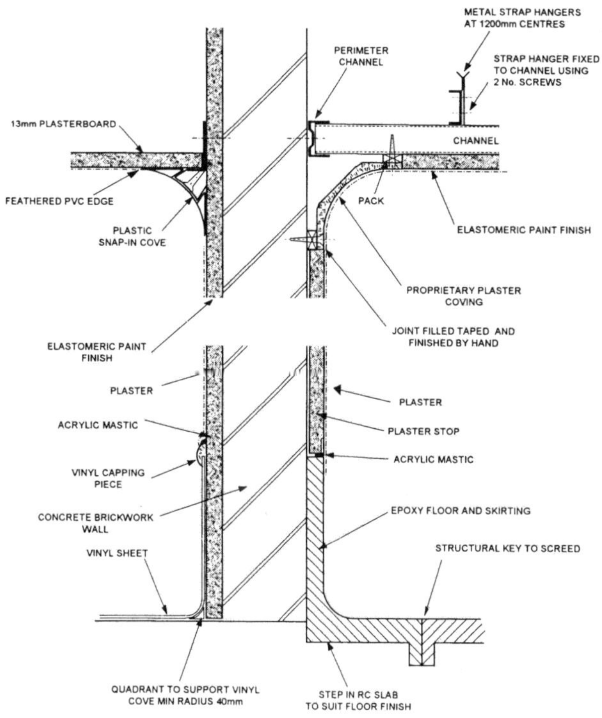

Figure 6.2. Typical coving details with *in situ* floors/gypsum board ceiling.

smooth, hard finish with rounded shapes for easy cleaning. Materials should be nylon-coated steel, chrome-plated or stainless steel.

4.4 Fabric Interfaces

Minimization of interfaces with the clean-room fabric should be considered and in particular the following points addressed:

- Services and distribution pipework increase the amount of surface area for gathering dust in clean rooms, and therefore increase the difficulties in cleaning and disinfection. Generally speaking, services distribution and utility pipework should be minimized inside the clean room, and should utilize adjacent but separate spaces or manifold rooms, permitting ease of maintenance.

- Where utilities and services are required to enter the clean room, the penetrations should be grouped together and manifold plates should be utilized, sealed against the room finishes.
- Maintenance, both routine and long-term repair or replacement, should be addressed and the requirements for access and equipment interchangeability incorporated into the design.

Figure 6.3 illustrates many of the key issues affecting the design of the clean-room floors, walls and ceilings and their detailed interfaces.

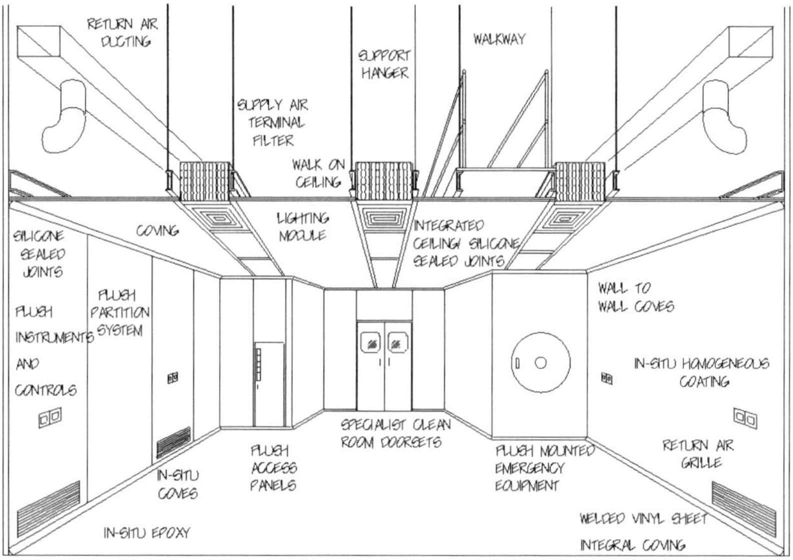

Figure 6.3. Key aspects of clean-room design.

5 APPLIED FINISHES

5.1 General

In selecting the materials for clean-room finishes, the installation costs should be considered against the maintainability of the material and ease of repair or replacement. The finishes must be able to accommodate the integration of the various fixtures and fittings.

Aspects of fire protection must also be considered and should at least address the issues of surface spread of flame- and fire-resistant construction.

Materials for the clean-room floors, walls and ceilings would generally be selected from the materials identified in Tables 6.1 and 6.2.

5.2 Floors

The performance of floors is a key issue and contributes significantly to the overall success of the clean room. All the following must be carefully considered when selecting the floor finish (with the order of priority determined by the individual project requirements)

- Chemical-proof
- Bacteriostatic
- Stable in dimension
- Colour fast (with suitable colours specified)*
- Good resistance to surface spread of flame
- Sound absorbing
- Antistatic
- Slip-resistant
- Resistant to abrasion
- Impact resistant
- Easily cleaned
- Colour-coded (between areas of different functions)
- "Soft" to walk on for operator comfort

5.3 Walls

Many of these attributes also apply to wall finishes, which also have an important function to fulfill by helping to contain the room environment. While they are more visible surfaces, they have to accommodate most of the penetrations into the clean room. The selection of finishes will be determined by the performance criteria required in terms of clean-room classification, robustness, and substrata construction, among others.

The requirements for cleaning, coved corners, integration with other finishes such as floor and ceiling surfaces, all have to be reviewed. The ability of the wall to accommodate clean details for penetrations and fixings must also be considered.

5.4 Ceilings

Ceilings can either be the underside of the upper floor, generally in concrete, or

*e.g., in oral solids plants, white walls should not be specified since contaminating dust cannot be seen.

Table 6.1 Clean-Room Construction Materials

Base Construction	Applied Finish	Commentary
Concrete	Power float or cement screed	Traditional wet trade construction, schedule impact and long drying-out time before finishing. Wide tolerances, requires subsequent finish. Robust construction but services integration has to be up front.
Brick/block walling	Render, plaster or gypsum board	Traditional wet trade construction apart from boards, schedule impact and long drying-out time before finishing. Robust construction but integration of services can be difficult.
Metal stud partition with gypsum board	Plaster, skim or taped joints	Lightweight construction, quick to erect and line out. Integration of services straightforward. Support of wall-mounted equipment has to be integrated early on. Costs about same as masonry. Allows freedom of choice of finishes.
Proprietary partition	Laminate or pre-finished metal panels	System designed wall panels easy and quick to erect with integrated joint details. Often combined with ceiling system to form complete clean room. Good standard of finish but tends to be at higher cost end.
Metal support system with gypsum board	Plaster, skim or taped joints	Traditional method of *in situ* ceiling, quick to install with freedom of choice of finishes. Above ceiling access can be limiting, although can be made walk-on.
Proprietary suspended ceiling	Pre-finished metal tiles	Wide range of sizes, finishes and joint details available from simple epoxy metal tiles exposed joints silicone sealed to sophisticated gasket joint systems.

Table 6.2 Clean-Room Finishes

Finish Material	Location	Comments	Comparative Unit Cost
Epoxy screed	Floor	Wet trade, time consuming in project schedule. Hard wearing, various grades/thickness available. Very/good chemical resistance. Difficult to repair. Subject to cracking due to shrinkage/settlement.	3
Polyurethane terrazzo screed	Floor	Wet trade, time consuming in project schedule. Hard wearing, various grades/thickness available. Very/good chemical resistance. Difficult to repair. Subject to cracking due to shrinkage/settlement.	9
Terrazzo	Floor	Tiles or in-situ, wet trade with schedule impact. Hard wearing, attractive appearance. Chemical resistance good. Tiles relatively easy to repair. Subject to cracking due to shrinkage/settlement.	5
Welded PVC/ vinyl sheet	Floor, wall ceiling	Base preparation important. Quick to lay and welded joints give good continuous surface finish. Limited resistance to chemicals and heavy traffic. Good flexibility and easy to repair. Relatively underfoot for long working hours. Not suitable for heavy turning vehicles. Requires proper preparation of substrate especially moisture content of slab	2
Epoxy paint	Wall and ceiling	Base preparation important. Quick to apply and gives good continuous surface. Good chemical resistance. Repair easy but use solvent free grades. Limited resistance to fabric movement.	1.5
Elastomeric paint	Wall and ceiling	Base preparation important. Quick to apply and gives good continuous surface finish. Limited resistance to chemicals. Good flexibility and easy to repair. Limited resistance to abrasion.	1
Glassfibre coating	Wall and ceiling	Base preparation important. Long application has schedule impact. Gives good continuous surface. Good chemical resistance. Repair easy with good resistance to fabric movement.	3
Glass reinforced plastic (GRP)	Wall and ceiling	Normally part of a proprietary system. Panel joint details important. Fair chemical resistance but difficult to repair.	5
PVC coated steel	Wall and ceiling	Normally part of a proprietary system. Panel joint details important. Fair chemical resistance but difficult to repair.	5
Phenolic resin sheet	Wall and ceiling	Normally part of a proprietary system. Panel joint details important. Good chemical resistance but difficult to repair.	5
Stainless steel	Wall and ceiling	Good at covering base imperfections. Sheet size and joints a disadvantage. Good chemical resistance	5
Enameled steel	Wall and ceiling	Normally part of a proprietary system. Panel joint details important. Fair chemical resistance but difficult to repair.	3.5

false, suspended from the floor above. If a false ceiling is provided then it is of primary importance to establish the method of maintaining ceiling fittings.

Two options are available.

- Bottom access from within the clean room, although this entails maintenance personnel entering the clean room and will more than likely require the room to be revalidated
- Alternatively, access can be designed from above the ceiling, via walkways; even a complete walk-on ceiling can be provided, with the implications for increased construction costs. Ceilings must be airtight and able to maintain any over- or underpressure that could be required in the room. There must also be dimensional compatibility between the ceiling, light fittings, HVAC grilles and other fittings, and they should all be detailed as flush as possible with the ceiling surface.

Ceilings will usually be one of the two following types.

- *In situ* suspended, usually a plasterboard system, with an applied coating finish
- Prefinished composite square or rectangular panels in an exposed or concealed supporting grid

A further alternative to this is the use of composite units, usually long rectangular panels with a tight tongue-and-grooved joint on their long edges. All of these systems can be detailed to be "walk-on."

5.5 Painting

The correct preparation of the surface before painting is critical to achieving a good finish. The surface must be smooth, free from any loose material, and have the correct minimum moisture content. The types of paint most commonly used are epoxy, elastomeric and acrylic. It is preferable that all paints are aqueous-based, since this eases the application, particularly in small rooms.

6 FIXTURES AND FITTINGS

6.1 Openings and Penetrations

Not only must openings in the clean-room walls for doors and vision panels be detailed flush with the adjacent walls, but they must also take account of the following criteria.

- The door seal against the doorframe should be flush detailed, utilize recessed sealing strips, and designed to a specified air-leakage rate. Currently one of the only recognised standards for air leakage across doors is BS 476 Part 22.
- The doorframe components should be minimized with integrated stops or architraves and present a smooth finish with rounded internal and external corners.
- All metal frames should have welded joints ground smooth.
- Wall vision panels should be flush glazed with any fire rating provided by a supplementary layer of glass within the depth of the wall.

6.2 Doors

- Clean-room doors must be of specialist design and formed from various materials, such as steel, glass reinforced plastic (GRP), glass and other laminate constructions.
- All doors should have a smooth, uniform surface without visible projections and irregularities.
- Joints in the door construction should be positioned on the vertical edges only.
- Specialist clean-room doors should be used and be prefinished where possible with hardware minimized and factory fitted.
- Core materials and finish should be carefully selected to meet the performance requirements (including fire rating if necessary) and door vision panels flush detailed and factory fitted.
- If hollow metal doorframes are used in masonry walls, then the voids should be filled with cement mortar as the wall construction proceeds.

Figure 6.4 illustrates the principles to be followed.

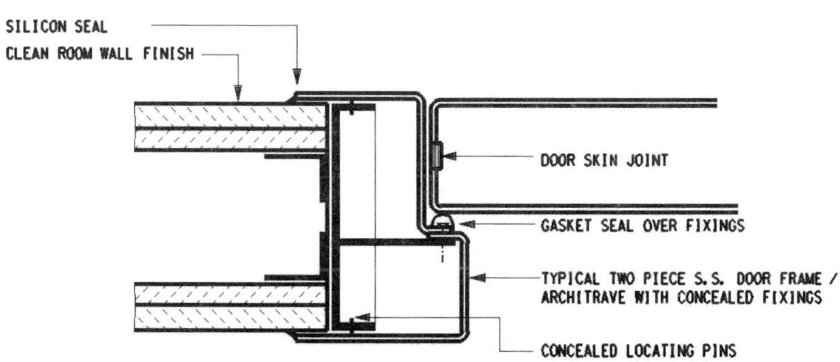

Figure 6.4. Typical flush-mounted clean-room door.

Glass doors should be manufactured from toughened glass, no less than 9-mm thick. These doors are best set in a stainless steel frame with stainless steel door furniture.

Emergency exit or fire-resistance doors must comply with these requirements and preference should be given to proprietary crash panels specifically designed for clean rooms.

It should be noted that all other penetrations into the clean room must be treated in a similar manner. This would include equipment penetrating through the walls, ceilings or fixed into the surface itself, and such items as utility or service panels, light fittings, HVAC grilles and filters, control panels, CCTV, speech panels, key pads, touch telephones, sprinkler heads or covers and emergency showers.

Other penetrations into the clean room must be sealed flush using gaskets, manifold plates or a silicone mastic. All joints between different materials, or where surfaces are not flush (maximum should be 5-mm to 10-mm projection), should be sealed with a silicone elastic seal, which must be smooth and have a minimum/maximum depth and width of 5 to 10 mm.

All silicone mastic used in the clean room should be formulated with an antibacterial additive and the gasket material should be a smooth-surfaced, closed cell-type rubber such as Ethylene Propylene Diene Monomer (EPDM), which provides a smooth, high-performance membrane.

6.3 Windows

If the clean room is situated on an outside wall and windows are provided, they should have steel frames and be flush with the walls on the inside. If this is not possible, they must have a sloping sill with a slope of at least 60 degrees finished in a suitable impervious material such as stainless steel. The windows should also take account of environmental issues such as thermal insulation, solar heat gain, noise and glare.

Internal windows, normally referred to as vision panels, should be flush detailed with the wall and, if situated between two clean rooms, the panel should be double-glazed to allow flush detailing on both sides. If double glass is used then regenerated silica gel (50 to 100 g) should be inserted in the void to avoid condensation.

Where fire resistance is required, the method of retaining the fire-rated glass with mechanical fixings requires that this glass be fixed centrally in the wall. Figure 6.5 illustrates the key aspects of a flush fire-rated vision panel.

6.4 Lighting Fixtures

Light fixtures should be specialist clean-room fittings, normally fitted flush to the ceiling. Some particular clean-room designs (e.g., teardrop fittings) can be surface

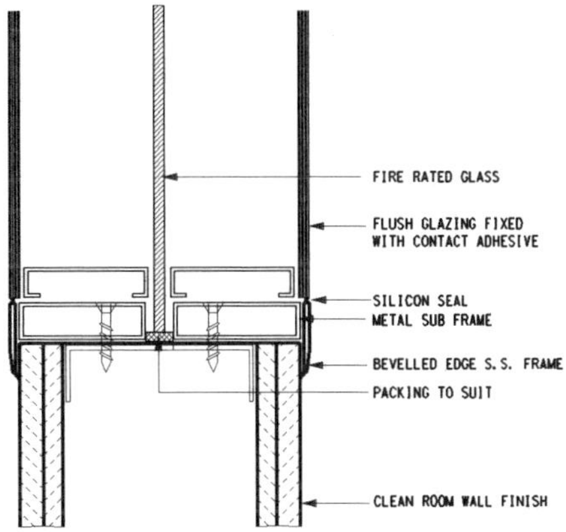

Figure 6.5. Typical integrated flush vision panel.

mounted in nominally clean areas, and be up to ISO Class 8 classification, provided they have a good seal between the base of the light fitting and the ceiling itself.

The diffusers should not be glass (for general safety issues) and must be smooth and easy to clean. The method of maintaining the light and replacing fluorescent tubes must also be considered, together with its implication on the ceiling design as noted in Section 5.4.

6.5 Sanitary Appliances

The only types of sanitary appliances found in clean rooms are production sinks, usually located in ISO Class 8 or lower clean rooms; hand-wash stations are usually located in changing areas or air locks. Appliances must be of a suitable design with clean lines, no sharp angles, and no recessed corners.

The materials of construction should be either stainless steel or ceramic, and all support fixings, water supply and waste pipework should be minimized.

Alternatively, many installations will adopt the philosophy of exposed frame sink supports with no cupboards. This encourages the avoidance of hidden traps, and lessens the likelihood of poor housekeeping due to the high visibility of the area. The joint between the appliance and any floor or wall surfaces should be clean, straight and filled with silicone mastic.

If taps are used to control water flow in hand-wash sinks, they should be elbow- or foot-operated, rather than by hand. Alternatively they may be actuated by photocells or other noncontact means.

Traps should be of the pop-up waste type, and fittings minimized where possible by using hospital-type mixer taps; or, preferably, actuated by an electronic sensor.

6.6 Floor Drains

Floor drains would not normally be specified in ISO Class 7 or higher categories of clean room, and as a general rule they should be minimized wherever possible. If unavoidable, they should be finished flush with the floor, constructed of stainless steel (grade 316 is normally acceptable), and be complete with a removable airtight solid cover. Proprietary clean-room types should be used that allow a designed integration with the floor finish and have an integral trap removable from above the floor via the access cover. Equipment installed in pharmaceutical clean rooms should not be plumbed directly into floor drains.

6.7 Services

Electrical fittings within the clean room should be minimized where possible by using remote switches or automatic censors for lighting control. Power outlets should be grouped and integrated within utility panels on the wall. The design of the fittings should be flat, with no sharp edges, recessed corners, and the fittings be complete with cover flaps. Their construction material could be plastic for up to ISO Class 8 areas, but for ISO Class 7 and higher classifications, stainless steel is recommended.

6.8 Fire and Building Codes

Construction must also address the requirements of any building codes and fire strategy adopted for the building containing the clean rooms. Such buildings often have a complex of rooms, linked with air locks and clean corridors. Fire protection of the structure and personnel escape routes must be compliant with statutory regulations and codes, and individual company standards. Specific periods of fire resistance and compartmentation of the building may be required to isolate areas of special risk or particularly hazardous materials or operations.

Very often, particular codes cannot be complied with. In these cases, detailed discussions and explanations must be given to the authorities. Typically

- Step-over benches in changing airlocks would not normally comply with safety codes if these were also designated as emergency escape routes
- In emergency escape routes, single swing doors opening against the direction of escape for operational or air pressure regime requirements would again not normally comply

In such cases, it is common to obtain a waiver or relaxation to allow the infringement. However, experience shows that this is never guaranteed and each individual application will be evaluated against the particular circumstances of the design.

It is worth noting that any extenuating circumstances, such as fully automatic alarms or sprinkler protection, can often be taken into account to compensate for any waiver requirement.

Alternatively, if the authorities respond negatively, then other options, such as proprietary sealed push-out panels, could be integrated into the clean-room walls. These panels are normally fitted with easily removable or pull-out retaining gaskets to facilitate the emergency escape.

Where possible fire extinguishers and hose reels should be outside *but near the entry* to the clean room. Where equipment has to be inside the clean room, then it must be contained in a recessed box, finished flush with the wall and fitted with a solid door of glass or metal to achieve a good airtight seal with the frame of the box.

Finally, the spread of fire, smoke and hot gasses must be controlled in any voids in clean-room walls or above clean-room ceilings must be controlled by suitable fire and smoke cavity barriers.

7 CLEANING AND CLEANING MATERIALS

7.1 Cleaning Methods

Methods of cleaning and disinfection must be recognized as key aspects in the selection of the finishes materials. The following should be carefully reviewed, and any decisions taken clearly recorded, in conjunction with the client and the end user:

- Cleaning by simply wipe-down and swabbing, but noting what particular cleaning agents are used
- Sanitizing the services and the use of a disinfectant agent for the wipe-down process
- Sterilization by the use of gassing or fogging, affecting the entire clean room. This sterilization will use agents such as hydrogen peroxide or formaldehyde (although the latter is less common now)

The specifier must be fully aware of the above and make careful reference to the reagents and methods used, while choosing the finishes. Some materials used in the cleaning process are highly toxic or corrosive, and can have a significant chemical reaction with some finishes materials.

The client or end user of the various cleaning agents may well have previous experience of the use of these materials, and be able to advise on their chemical properties.

If available, this information should be referred to when making the final selections for the finishes materials. If unavailable, then manufacturers should be able to provide details of the performance of their materials with various chemicals.

If the specific chemicals used in the cleaning process are not known, then the manufacturer should be able to carry out tests on possible reactions, and identify the suitability of the materials under consideration.

In summary, the specifier must follow clear steps to ensure compatibility of cleaning materials and room finishes.

- Know the methods of cleaning and any chemicals that will be used
- Make preliminary selection of finishes materials, based on the above information and with reference to manufacturers' published chemical properties and resistance of the selected materials
- Complete testing of any unknown chemical reactions in conjunction with the manufacturer, client or end user
- Make final selection based on the above with documented back-up performance data

8 HVAC INTEGRATION

8.1 Air Distribution

The air distribution within the clean room must provide the room with the required supply and extract airflow rates, while having minimal impact on the room itself.

The method of air supply and extract chosen will depend on the type and grade (class) of clean room required. Low-grade rooms will tend to have air supplied at a high level via ceiling diffusers and low-level extract through sidewall grilles. The aim is to thoroughly mix the room air with incoming clean air to achieve the required air particulate level by dilution.

Higher-grade rooms requiring unidirectional airflow require more complex arrangements. Ideally rooms requiring horizontal flow should have the supply air delivered via an "air wall," complete with terminal filters and extracted via perforated panels in the opposite wall.

Rooms requiring vertical flow have the supply air delivered via a proprietary clean-room ceiling complete with supply air filters and ideally extracted via a false floor plenum.

However, false floors are unacceptable in some applications, e.g., pharmaceutical process rooms, due to the problems they pose with cleaning, and a compromise is needed.

Normally, low-level extract is provided, ideally by means of a continuous slot, or row, of extract grilles. As a minimum requirement these slots or grilles should be positioned in the two longest opposite walls of a rectangular room. It should be noted that, if the distance between these walls is greater than four metres, there will be a tendency to pull the air flow toward the walls (out of the vertical) at the working plane level, which may well be unacceptable.

8.2 Detailed Integration

The main challenge in keeping air distribution ductwork out of the clean room is to pick up the low-level extract points. For single-room applications with available adjacent space, there is obviously no problem, but for adjacent multiroom applications, several vertical drop ducts need to be incorporated in the walls or corners for low-grade rooms, and proprietary "air walls" provided for rooms requiring unidirectional air flow.

For unidirectional air-flow clean rooms, the air-flow patterns must be reviewed against the equipment layout and operator positions, particularly in sterile operations, to ensure that the specified patterns are achieved with no "dead spots"; and that the product and operators are protected from contamination.

The final integration of HVAC with the clean-room fabric occurs with the flanges of the HVAC grilles or terminal HEPA filters and the room surfaces. Generally, it is accepted that grille flanges will overlap onto the wall surface with a maximum 5 to 10-mm projection sealed against the surface with concave silicone sealant mastic. In proprietary wall and ceiling systems, all such fittings should be designed to be completely flush wherever possible.

REFERENCES

ISPE. *Pharmaceutical Engineering Guides for New and Renovated Facilities Bulk Manufacturing Facilities*, 1996.

ISPE, *Sterile Manufacturing Facilities*, 1999.

Kozicki, M.N., Hoenig, S.A., Robinson, P.A. *Cleanrooms — Facilities and Practices,* Van Nostrand Reinhold, New York, 1991.

Ljungvist, B., Reinmuller, B. *Clean Room Design,* Interpharm Press, Buffalo Grove, IL, 1997.

Schneider, R.K. *Practical Cleanroom Design,* Business News Publishing Company, Troy, MI, 1995.

The Institute of Quality Assurance Pharmaceutical Engineering Guides for New and Renovated Facilities, *Pharmaceutical Premises and Environment*, 1987 (revised 1997).

W. Whyte (ed). *Cleanroom Design,* Wiley, Chichester, U.K., 1991.

Chapter 7

Rapid Microbiological Methods Explained

Stewart Green and Christopher Randell

CONTENTS

0-849-32300-2/04/$0.00+$1.50

INTRODUCTION

Microbiology can arguably be considered as the original of the classical sciences. Although our ability to visualize, enumerate, and isolate most microorganisms is relatively modern, i.e., over the last 300 to 400 years, their impact was surely known to the earliest inhabitants of Earth. Putrefaction of meat and vegetable matter, while caused by a virtual microcosm of different genera, owes much to the activity of bacteria and fungi. Similarly, although the actual agents associated with specific diseases were not identified until the mid-1800s, it did not stop thoughtful men from musing on the agents of disease. Lucretius (95–55 B.C.) recognized the existence of "seeds" of disease.

While the other sciences have seen a veritable explosion of techniques and associated instrumentation over the last 200 years, microbiology techniques are often still based on work done in the late 1700s to mid 1800s, the vast majority of microbial enumeration is still carried out by the pour plate method developed in 1870 by Robert Koch.

Our guidance provides a review of techniques available to microbiologists to considerably improve both the accuracy and speed of their determinations. Unfortunately the take-up of some of these techniques has been slow for multifarious reasons that include regulatory attitudes in this arena, and the sensitivity of the areas impacted by microbiology in the pharmaceutical industry, such as the manufacture of aseptic sterile products.

Our purpose is to present a rapid overview of the most commonly available commercial rapid microbiological test methods. Our coverage includes a description of each method, with a proposal for where it can be used. Where appropriate, suggested "validation" provides guidance whereby users can achieve regulatory approval for use of their chosen method. It is not intended to be a treatise on microbiology. Refer to more detailed standard reference works for this purpose, at the end of this chapter.

This section discusses the principles of several rapid microbiological methods and their application within the pharmaceutical and medical-device industries. The food industry has already embraced many of these techniques.

The regulatory acceptance of these methods from a European and U.S. perspective is based on experience, and on public comments made with respect to the methods, by both the Medicines Control Agency (MCA) (now Medicines and Health Care Products Regulatory Agency (MHRA)) in the U.K., and the Food and Drug Administration (FDA) in the U.S.

The text is based on the authors' joint experience of utilizing the methods to deal with common pharmaceutical dosage forms, i.e., tablets, capsules, liquids, ointments, creams, and suppositories. Sterile products are only touched upon, as the authors do not have personal knowledge of the use of rapid methods in this arena.

Although a number of methods are examined, we concentrate on those suited to the hurly-burly daily activities conducted in a pharmaceutical or medical devices control laboratory.

While the nomenclature "rapid" is used to describe the methods evaluated in this text, it is perhaps better to consider them as "modern" as a more apt counterpoint to "traditional" or "conventional" methods.

The suppliers of media, reagents, etc., have made considerable strides associated with traditional methods, helping to improve their discriminatory ability, and the rapidity at which results are produced.

1 OVERVIEW OF "TRADITIONAL" MICROBIOLOGICAL METHODS

Depending on the nature of the product in which microorganisms are to be enumerated or isolated, there are a number of "traditional" long-standing techniques. Many of these are firmly established in the seminal work of the pioneers in the field during the 1800s. The methods are still valid in the traditional sense today and used in countless laboratories throughout the world, providing a relatively simple means of assessing the number and type of microorganisms in pharmaceutical products, or in the environment in which they are manufactured.

There are, however, a number of drawbacks, some of which are shared by rapid methods.

- For the purposes of enumeration, it is assumed that one visible colony is derived from one bacterial cell. This may lead to underestimation by a factor of ×10 to ×100, depending on the nature of the organism and the environment from which it is isolated.
- The approach is very much "one size fits all." It is assumed that the media selected will recover most types of organisms under the given incubation temperatures. Allowance is sometimes made for sublethally damaged cells by including a resuscitation step, but this may preclude enumeration.

- Different enumeration methods can give vastly different counts, which can be impacted by the technician's skill as well as media and incubation variables.
- They are time-consuming. From isolation to identification may take 7 to 10 days (or longer) for more esoteric isolates.
- They may lack sensitivity due, for example, to the high dilutions needed to remove the inherent inhibitory effects of the product.
- The product itself may make it difficult to discriminate between bacterial or fungal growth and dispersed product.
- Considerable infrastructure is required to support the methods, e.g., autoclaves, media steamers, laminar flow or biosafety cabinets, separate laboratory areas etc., an infrastructure making cost comparisons between traditional and rapid methods difficult.

Notwithstanding these issues, the techniques have stood the test of time and many billions of pharmaceutical doses have been released to the market using such techniques to confirm the absence of objectionable microorganisms.

Traditional techniques most commonly used are:

- Pour plates — the product or a dilution of the product is carefully mixed with molten agar media at approximately 45°C or below, poured into Petri dishes, allowed to solidify, and incubated at appropriate times and temperatures. An additional drawback of this method is the effect of agar medium at approximately 45°C on sublethally damaged microorganisms.
- Spread plating — normally used for difficult-to-homogenize materials such as fats or where fungi are to be enumerated. Samples or a dilution thereof are simply uniformly spread over the surface of a solidified agar medium and then incubated. While this overcomes the impact of warm agar on sublethally damaged cells it has the drawback of little or no dilution of any product inhibitory impact, such that bacteria or fungi may be present but, because they do not proliferate, cannot be counted. Spreading over the surface uniformly is a difficult technique, leading to underestimation of the counts due to confluent growth.
- Drop counts (Miles Misra) — here a low dilution of highly soluble or aqueous formulations is applied to the surface of a predried plate using a calibrated dropper, the plate is incubated and the number of colonies per drop counted. The principal disadvantage is its low sensitivity, which limits it to product containing more than approximately 500 colonies per ml, and the small area covered by the absorbed drop that results in counts of no more than 30 to 50 per drop, over which count confluent growth and under estimation is the outcome.
- Most probable number (MPN) — basically the dilution of a sample in a range of 1 in 10 to 1 in 1000 added to triplicate 9-ml volumes of tryptone soya broth. Using statistical tables and the number of tubes at each dilution step showing growth after incubation, the number of bacteria can be estimated. The main

drawbacks are the sensitivity (published tables provide, in theory, the ability to enumerate down to 3 cfu; however, as this is only at 95% confidence then the true result could be as high as 17); and that the whole technique is based on the assumption that microorganisms are normally distributed in a sample, although this is not proven on a case-by-case basis.

- Direct or microscopic counts — as the title implies, this is the examination of the product or dilutions thereof in a counting chamber under a microscope. The drawback here is that the technique requires considerable microscopy skills to establish good clear solutions.
- Membrane filtration — the product where possible or dilutions thereof are filtered through a bacterially retentive membrane, which is then washed and aseptically transferred either to broth for presence or absence, or agar for enumeration following incubation. Drawbacks are the expense of the apparatus, the need for a laminar flow unit (although this can be overcome by the use of closed, disposable membrane filtration units), the effect the product may have on the membrane rendering it porous or causing degradation, and its limitations when trying to filter fatty or very viscous solutions. Care must also be taken when a vacuum is applied to assist the filtration, that the membrane is not excessively dried, leading to death of microorganisms due to desiccation. Current practice is to use a membrane of porosity of 0.45 micron, though there are bacteria known to be capable of passing such filters; either because they are habitually small, or the substrate in which they are growing exerts pressure on their ability to grow to normal (i.e., greater than 0.45 micron) size. This may lead to *underestimation of the count.*

One further limitation common to all the above methods is their ability to be validated when compared with modern standards applied to other analytical techniques. This can often lead to unfair comparisons drawn by the regulatory bodies when assessing the substitution of a traditional method with a rapid one.

2 RAPID MICROBIOLOGICAL METHODS IN PRACTICE

Numerous microbiological methods can be considered "rapid" in comparison with traditional ones. These range from techniques that can give results within a few minutes, to those that give results within approximately 24 hours. Not all may be commercially available; some are more suited to research facilities, because they are either too complex or insufficiently robust for routine use within a quality-control environment.

Before contemplating the use of a rapid method it is worth quickly reviewing why a rapid method might be selected. The pharmaceutical and medical-device industries traditionally achieved their competitive edge by the innovation of their product portfolio. While this is still the case, it has become increasingly more difficult to identify novel therapeutic compounds, and more expensive to bring to market. It is said that only one in every 10,000 compounds evaluated are ever commercialized, at a cost of between $250 to $500 million to the pharmaceutical industry.

Companies have therefore had to look at other areas to maintain profitability, including mergers or takeovers, centres of excellence, product portfolio rationalization, and customer service. It is the latter area that has promoted the use of rapid methods.

While most chemical analysis can produce results ranging from a few minutes to 24 hours, microbiological analysis takes between two to seven days (up to 14 days for sterility testing). Therefore rapid method introduction can significantly increase speed to market, release inventory faster, reduce warehouse storage capacity, and enable the market to be serviced more rapidly. Against this capital cost of the equipment required, and, invariably, the higher cost of associated consumables must be offset. However, where the product impacted is becoming increasingly more expensive, these costs can normally be amortized against the faster turnover of the inventory.

We have identified the principal rapid methods proposed for either identification or enumeration of microorganisms. Where the technique is either to the authors' knowledge not commercially available or suited to routine use, this is noted in the accompanying text. Identification methods specific for a single microorganism have not been included.

2.1 Identification Methods

Such techniques can be applied in a number of pharmaceutical industry situations. For example, isolates from products or water systems, particularly water for injection systems, provide an ideal opportunity to use real-time methods. Isolates from within sterile areas can, if rapidly identified, provide a greater opportunity to determine the potential source. In all cases prior to the use of a rapid technique to effect the identification, a pure (i.e., single species) culture must be prepared using traditional techniques; hence the advantage of speed can be lost. What is gained is the accuracy of the identification — accuracy of approximately 80% can be expected for most techniques.

2.2 Miniaturized Detection Kits

Miniaturized detection–identification kits such as API, Enterotube, Vitek, Biolog and B D Crystals are typical examples. All work on basically the same principle: a pure isolate of a bacterium is Gram stained and subjected to a small range of rapid chemical tests, e.g., catalase, oxidase or coagulase. This enables selection of the correct test kit type. A culture is then suspended in saline and added to a succession of wells, which contain a variety of growth substrates with or without indicators: the culture is then incubated. As the organism grows and ferments and utilizes the substrate, the by-products of its metabolism induce a colour change in the reagent. The combination of changes is compared either manually or automatically against a database derived from profiling literally hundreds of different bacteria isolates. This hopefully relatively unique "fingerprint" of reactions allows an identification to be made.

In many cases the system is microprocessor driven and a "fit" to the profile is also given, providing a level of assurance of an accurate identification. Most of these systems provide an identification between four and 24 hours, considerably faster than conventional methods.

2.3 Fatty Acid Analysis Using Gas Chromatography

The fatty acid content of bacterial cells is relatively constant within a taxonomic group, therefore utilizing gas chromatography (GC) allows the identification of individual genera. However, this technique requires the use of sophisticated equipment, lengthy preparation of the cultures to enable their analysis, and careful control of media composition as this will impact on the speciation.

Most systems are computer-controlled and are able to generate individual profiles for common isolates in a particular environment. Available commercial systems are expensive and may not be considered robust for routine use.

2.4 Electrophoresis

In this technique the culture is grown in the presence of a radiolabeled protein for a short period, and is then incorporated into the cells. A suspension of the culture is applied to an electrophoresis plate, and a high voltage applied to separate the cell protein into a band, which can then be visualized by exposure to x-ray film. This banding is unique to each species of microorganism.

The authors are unaware of any commercial application of the method, which also requires specialized techniques and equipment. The reader is referred to *Proteomics Review 2001* by Michael Durin for further information

2.5 Mass Spectroscopy (MALDI-TOF)

Matrix-assisted laser desorption ionization time of flight (MALDI-TOF) mass spectrometry is a very sophisticated technique with roots in the biotechnology industry. In essence, if a cell culture is volatilized, then ionized, the resulting ions can be accelerated in an electrical field to provide a beam. This beam is then deflected using a magnetic field. The degree of deflection is dependent on the ratio of the mass of each ion to its charge, those with the largest ratio being deflected the least. The beam hits a slit, which literally takes a slice of the ion beam, which is then detected.

It is the pattern of these charged ions that can be used to identify at least to genera level, and in some cases to species. While apparently highly discriminatory, the technique cannot be considered suitable for routine use.

2.6 ELISA (Enzyme-Linked Immunosorbent Assay)

As is the case for the previously described MALDI-TOF technique, ELISA techniques also had their roots in the immunology and biotechnology fields. Again this is a complicated technique, requiring the generation of specific antibodies to an organism. A second antibody linked to an enzyme is also prepared, and is specific to the primary antibody. This enzyme has the ability to convert a colorless substrate to a colored one. The organism (antigen) is isolated and fixed to a substrate and the primary antibody applied. If it is a "match" it will adhere to the organism. If the antibody were not a match, by washing the substrate, it would be removed.

The secondary antibody is then applied and if the organism has been matched it will attach to the primary antibody. The substrate is washed again and the indicator applied. The enzyme on the secondary antibody cleaves the indicator to produce a visible colored compound. Clearly, if there is no match at any stage, the colour does not develop. A large number of specific antibodies need to be available, therefore this technique is often used after an initial screening has narrowed the choices. Once again this would not be a technique suitable for routine use.

2.7 Nucleic Acid Technique

The bourgeoning biotechnology industry has developed a number of techniques based on looking at the nucleic acid in the cell. The principals employed in the technique are similar whether a desoxyribonucleic acid (DNA) or ribose nucleic acid (RNA) probe is used. A culture is subjected to heat to denature it, and a specific single-stranded DNA probe is introduced, which binds to a target on the cell DNA; this double-stranded section is detected by a suitable label. A commercial RNA probe kit is available that uses an enzyme to cut up the bacterial DNA, and an RNA probe is used to target the fragments. In both techniques amplification (i.e., the production of numerous copies using techniques such as polymerase chain reaction (PCR)), may also be utilized to increase the response. It is the pattern of hybridized nucleic acid that is used as the means of identification. This technique can produce results in hours but does currently require a different skills set from the microbiology department. However, as commercial apparatus is further developed, the technique will be reduced to "button pushing" and arguably will sit more readily in the chemical QC section.

2.8 Fluorescent Probes

This is an adaptation of the above technique in which a fluorescent marker is added to either an antibody or a nucleic acid probe, such that when the probe locks with the target site — a cell wall, polysaccharide, nucleic acid, etc. — it can be visualized using a fluorescence detector. As for other such techniques because it relies on a targeted approach, it is considered more useful when the presence of a specific organism is anticipated, for example, the detection of pathogens.

2.9 Fourier Transform Infrared Spectroscopy (FTIR)

FTIR has transformed the identification of chemical raw materials in the pharmaceutical industry and shows promise for the quantitative analysis of the finished product. The basis of the identification is that different molecules are excited by absorption of infrared radiation, and this excitation can be measured and a fingerprint of absorption maxima determined. Bacteria can be subject to the same technique and produce highly complex absorption patterns. By selecting a specific spectral range across which to measure the fingerprint, some of this complexity can be reduced but still enable, by comparison to the spectra of a known organism, an identification to be made. Although experimental data is scarce the technique appears to even differentiate between different strains of the same organism, at least across the limited range studied. The authors are not aware of any commercialization of the technique but clearly with the widespread use of FTIR for chemical analysis, this should not prevent its use.*

3 ENUMERATION AND PRESENCE METHODS

It is the arena of rapid enumeration and detection methods that probably holds most appeal for pharmaceutical companies. Many companies are driven to achieve release from a microbiological perspective within the same timescale as for chemical analysis, i.e., 24 to 48 hours rather than the five to seven days of a traditional microbiological method.

The four main areas for this drive are:

- Raw materials testing (natural origin)

---

*Further details of the technique may be found at www.aiha.org.

- Finished product testing
- Sterility testing
- Environmental monitoring swabs or air samples from clean rooms

Certainly we have personal experience of gaining widespread regulatory approval throughout the E.U. for finished product testing and are aware that approval has also been achieved in the U.S. Although theoretically rapid techniques could be applied to the sterility test, the sensitivity of most of the methods is proving a stumbling block, with the best systems still requiring the presence of 5 to 10 organisms per 100 ml sample. However, development of these systems is ongoing and it is surely only a matter of time before the 14-day sterility test could be supplanted by the one-day test!

3.1 ATP (Adenosine Triphosphate)-Based Systems

The reaction below has been known since the 1940s and has been commercially exploited in its current format since the 1980s. Despite this relatively long history the pharmaceutical industry has been conservative in its uptake of the technique, even in the face of qualified encouragement, from some of the regulatory bodies.

$$\text{ATP} + \text{D-luceferin} + O_2 \xrightarrow{\text{Luciferase} + Mg^{2+}} \text{AMP} + \text{oxyluciferin} + CO_2 + \text{Ppi} + \text{light}$$

- The enzyme luciferase hydrolyses ATP in the presence of oxygen and magnesium to produce light at a wavelength of approximately 562 nm. The amount of light emitted is directly proportional to the amount of ATP present, ATP only being contained within living cells. As the amount of ATP in bacterial and fungal cells is relatively constant (regardless of genus/species) at 1.0 to 1.5×10^{-15}g for the former and 1.0 to 1.2×10^{-14}g for the latter, then the number of organisms present can be determined from the number of "light units" emitted during the above reaction.

In use, the sample in which organisms to be detected is treated to remove non-bacterial and fungal sources of ATP, preincubated to increase any organisms present, the ATP released from the cells, then the reaction initiated. The pre-incubation step can be just a few hours, but this obviously means that the number of organisms cannot be quantified, only their presence or absence determined. (This technique also offers the opportunity to subsequently identify organisms present following enumeration, as it can be considered nondestructive.) This technique is limited to pharmaceutical products with minimal to zero bioburden.

Literature would suggest that due to the efficiency of the light detection, the sensitivity of the method could be limited to the detection of not less than 100 cfu/ml. However, in our experience advances in the commercial application of the technique have increased this sensitivity to 2 to 10 cfus, and enumeration down to 1 cfu can be achieved .

The Steriscreen, Microcount and Pallcheck systems utilize this technique and the former has been extensively applied by one of the authors (C. Randell) to a wide range of pharmaceutical products including liquids, creams, ointments and suppositories with equal success. We have achieved regulatory approval in a number of E.U. countries, substituting the Steriscreen for the traditional method, with the caveat that, should microorganisms be detected, they would be quantified and identified where necessary using traditional methods. The time taken to achieve a result is generally 24 hours for bacteria and 48 hours for yeasts or fungi. The method has proved robust and reliable and very few products containing high ATP from nonmicrobial sources required further processing to remove the effect.

3.2 Microcalorimetry

Since the 1980s it has been observed that the catabolic activities of microorganisms could be measured using sensitive calorimeters. The technique demonstrated that the thermal profiles of different organisms were dissimilar. Early consideration of the technique as a means of identifying microorganisms did not materialize. The major drawback of the technique is that, even using the most sensitive calorimeters, counts in the order of 10^4 are required, thus limiting its usefulness.

3.3 Impedance

Impedance, like microcalorimetry, measures the changes in the growth media due to the metabolic activities of the contained microorganisms. In particular, the breakdown of large weakly charged molecules, such as proteins, results in the formation of many strongly charged amino acid molecules. This shift in ionic strength can be indirectly measured by the resistance in the growth media to the passage of an electric current. The relationship is defined by the equation:

$$Z = \sqrt{}R^2 + (1/2\pi fx)^2,$$

where

Z = impedance
R = resistance
c = capacitance
f = frequency

In application the electrical signal of a culture is continuously monitored, and at a certain level, when the results of the organisms metabolism allows the conductance of the electricity to be measured — the so-called detection threshold — the presence of this growth is detected. The technique has been commercialized and is extensively used in the food industry where microbial levels and limits can be considerably higher than in the pharma industry.

The technique has a number of drawbacks. The media used is very specific and is not that featured in the major pharmacopoeia. The high limit of detection demands preincubation for most pharmaceuticals, hence the system can only be used for detection, not enumeration. Similarly, the time for very low levels of microorganisms to reach the detection threshold in what could be a hostile environment, can be extensive. Finally, a number of organisms (notably non-fermentative Gram-negative bacteria) do not produce significant changes in the electrical characteristics of the growth medium, again leading to long incubation periods. Despite these drawbacks, some success has been achieved in using impedance for preservative efficacy screening, although the authors are not aware that regulatory approval has been gained for this application.

3.4 Direct Epifluorescent Filtration Technique (DEFT)

This technique has also been extensively used in the food industry. It utilizes the observation that viable cells, when exposed to acridine orange, can be visualized under a fluorescent microscope, appearing bright orange, while nonviable cells appear green. By sample filtration, the viable cells can be stained *in situ* and immediately counted, giving a result within one hour or less. The most obvious drawback is that the technique is limited to filterable samples and its sensitivity, to some extent, determined by how much of a sample can be filtered. Although some workers have detected down to 10 to 20 organisms per ml by filtering large volumes (litre quantities), generally the detection limit is of the order of 10^2–10^4 organism/ml. Other issues are the occasional nonselectivity of acridine orange, which can stain certain nonviable cells, and has a tendency to also stain other noncellular material in a sample.

However, because of the rapidity of the method, considerable development of the fundamental technique has taken place. The commercial equipment manufacturers Chemunex (an international company with its head offices in Paris, France) have developed equipment that overcomes at least two of the major drawbacks of the acridine orange methodology.

The first is the utilization of a more specific "stain" based on fluorescein. Samples are filtered, and then treated with the stain. Once taken into the cell, the stain is cleaved by an esterase releasing a fluorochrome that can be detected by laser scanning. Additionally, only cells with an intact cell membrane (i.e., viable cells) can retain sufficient amounts of the stain to be detected. Following laser scanning, the associated software logs every fluorochrome "hit," enabling true enumeration to be achieved since it has been "taught" to ignore fluorescing debris below a predetermined threshold. At worst this would probably lead to falsely high counts. The use of the laser and software also removes the labour-intensive microscopic examination required by the acridine orange method.

One area where the Chemscan RDI system has been used is in the monitoring of pharmaceutical water systems, where the provision of real-time results can potentially prevent the use of out-of-specification (OOS) water being used in the compounding of an expensive active. An interesting corollary is that the system appears to result in higher counts than traditional methods. This may be due to normal culture media offering less than optimal recovery for all organisms, or possibly enumeration of so-called viable nonculturable organisms by the Chemscan system. This poses a regulatory dilemma, in that although the water quality is no worse than it has ever been, it is possible that it will fail the pharmacopoeial standards. This dilemma has yet to be resolved in all countries but the MCA (now MHRA) at least recognizes the issue and accepts that counts may be higher without "failing" pharmacopoeial standards.

A further development of the fluorochrome labelling system is its combination with flow cytometry. The latter technique has been around since the early 1960s and basically consists of passing microorganisms, diluted in an electrolyte, through a very small aperture across which an electric current is applied. As the microorganisms pass through the aperture the electrical resistance changes can be measured. The disadvantages of this technique are that many other substances can alter the resistance, and the aperture can easily become blocked. The so-called D-Count apparatus uses the fluorochrome to label viable cells, which are then flowed past a laser beam that excites the fluorochrome, subsequently measuring it using a photomultiplier. As yet the sensitivity at reportedly 50 to 100 cfu/ml is still an issue, although recently sensitivity claims down to 1 cfu/ml have been made.

As might be expected from such a sophisticated computer-based system, the initial capital outlay for such systems is an issue. In today's climate, the more sophisticated the systems, particularly when involving computer software and hardware, the more sophisticated and time-consuming the equipment qualification will be, such as conformance to Good Automated Manufacturing Practice (GAMP) and CFR 211 Part 11 regulatory compliance. To our knowledge this technique has now gained regulatory approval for process water testing, and release testing for nonsterile products in Europe.

4 THE INTERNATIONAL REGULATORY POSITION

The authors have been successful in achieving regulatory acceptance by a number of Boards of Health within the E.U. for the substitution of a rapid method based on ATP/luciferase for traditional methods. However, this acceptance does not appear to be shared by regulatory authorities worldwide. At an advisory meeting on rapid methods within the FDA in May 2002, concerns were raised about:

- The ability of rapid methods to distinguish between viable and nonviable organisms. Equally, with traditional methods the assumption is made that the media, time and temperature combination selected is suitable for the enumeration of all microorganisms regardless of *genera* or state of health
- Increased sensitivity such that a rapid method may provide results that are outside compendial limits and would necessitate a relaxation of the limits for certain products. This ignores the fact that rapid methods are simply detecting what has always been there, and which presumably has not given rise to any patient-safety issues. It also ignores the positive benefits that manufacturers armed with this information can more critically evaluate their process but now with the opportunity to do so close to "real time," to see if such counts can be reduced, i.e., an improvement in product quality

Certainly within the U.K., the MHRA recognizes the challenge of rapid and generally more sensitive methods and is on record as stating that "for water used in pharmaceutical production, limits may have to be adjusted to compensate for this increased sensitivity."

The major pharmacopoeias have always recognized that methods other than those specified may be used, always with the caveat that where differences become apparent or in the case of dispute the pharmacopoeial methods will prevail.

The *USP* states: "Compliance may be determined also by the use of alternative methods, chosen for advantages in accuracy, sensitivity, precision, selectivity or adaptability to automation ... such alternative or automated procedures or methods shall be validated."

The *USP* also provides some limited guidance on validating microbial recovery methods (<1227>), pointing out the inadequacies of the plate count method in accurately enumerating counts, e.g., for counts of between 1 to 10 per plate, the estimated error of the mean is between 100 to 32%. Both the *USP* and the *PhEur* have a process for amending the pharmacopoeia. The *USP* has tabled at least two so-called "stimuli to the revision process" for rapid methods.

In summary, there still appears to be some reticence in the regulatory authorities to embrace new microbiological methods. In the U.S., this situation will probably resolve as alternative method validation is enshrined in the *USP*. In the E.U., manufacturers need to use the mechanism already available to them, i.e., the Type 1 marketing authorization process supported by an expert report to gain acceptance of rapid methods.

5 VALIDATION

It is worth reflecting that most of the general microbiological methods currently in use for total viable counts or recovery of named organism have not been "validated" as we understand the term today. Certainly they have a long period of use, but as debates on viable nonculturables, whether to use minimal media for recovery of water borne organisms, etc., have shown, this does not necessarily mean that the counts achieved have been accurate. Similarly, extensive work in clinical microbiology has repeatedly demonstrated that recovery of microorganisms from clinical specimens using standard isolation techniques is a far from robust procedure. Although the regulatory bodies insist on formalized validation, this must be set in the context of the widespread use of current methods, which can prove unreliable even in the most experienced hands.

For both qualitative and quantitative methods it will be assumed that the equipment used follows the traditional qualification process. Due to its very specific nature, apart from at a minimal level, it is probably unnecessary to develop a user requirement specification (URS). However, installation qualification (IQ), operational qualification (OQ) and performance qualification (PQ) are all required. As most of the equipment concerned is "driven" by computer software or hardware, the provisions of the GAMP guidelines published by ISPE should also be considered. This should certainly include:

- An audit of the software provider
- Access to the algorithms (or at least an agreement to how these might be accessed in the event of the company being dissolved for example)
- Life-cycle development; control of change, etc.

During the actual functionality checks it will be demonstrated that any security functions are operational; any additions, deletions or amendments to the data are recorded and ascribed to the individual performing the action, and the programme functions in a reliable and predictable manner as the menu is stepped through. Specific aspects relating to the equipment or method interface are now covered.

For the method itself, there has been growing support to treat microbiological methods where appropriate in the same manner as for chemical analysis. This approach was adopted to

- Meet the regulatory demands and
- Utilize an existing framework of tests by which the success or otherwise of the validation can be judged

Within this chapter, the definitions provided in the ICH/USP proposals have been used thus:

- Accuracy — the closeness of test results obtained to the true value of the article under test.
- Precision — the degree of agreement among individual test results when the analytical method is repeatedly applied to multiple samples. This may be measured either as reproducibility, i.e., when samples are analyzed at different times in different laboratories using different analysts or equipment or as

repeatability, i.e., the analysis over a short time period of a test article by the same analyst using the same instrument (sometimes also called ruggedness).
- Linearity — the test method's ability to provide test results that are directly (or by an accepted mathematical transformation) proportional to the concentration of the analyte in a sample of a given range (normally 50 to 200% of the expected value). The range is the interval between the upper and lower limits.
- Limit of quantitation — the lowest level that can be determined with acceptable precision and accuracy.
- Specificity — the ability to measure accurately and specifically the analyte in the presence of other components that may be expected in the sample matrix.
- Robustness — the ability of an analytical procedure to remain unaffected by small but deliberate variations in the test methodology.

6 IDENTIFICATION METHODS

Any of the previous methods will be supported by extensive databases from the manufacturers, which will include the results of probably thousands of determinations for the selected genera. It is unlikely that any such extensive cross-comparison could be performed by the user.

However, the methods may be based on completely different principles, e.g., biochemical metabolism, DNA probes, GC analysis of fatty acids etc., to methods currently being used. Hence the identification provided may differ at the species level and possibly at the genus level.

For most of these methods validation is limited to the following factors:

- Accuracy
- Precision
- Ruggedness
- Robustness

The normal microorganisms identified in a particular application should be considered. For example, if a manufacturer is constantly identifying Gram-negative nonfermenting bacteria, then more organisms representing this category should be included in the validation.

6.1 Accuracy

Two phases to accuracy determination are suggested.

First, type culture collections (ATCC or NCTC) for all those organisms routinely proposed for fertility testing by the major pharmacopoeias, should be determined using the existing method and the proposed rapid identification method. Normally this would embrace *Staphylococcus aureus* (NCTC 10788); *Bacillus subtilis* (NCIMB 8054) *Pseudomonas aeruginosa* (NCIMB 8626); *Clostridium sporogenes* (ATCC 19404); *Candida albicans* (ATCC 10231) and *Aspergillus niger* (ATCC 16404).

Second, recent isolates from the relevant environment applicable to the user should also be identified using both methods. Clearly the acceptance criteria is a consistent identification of the compendial-type cultures.

6.2 Precision

Using type cultures, this entails performing an identification on multiple samples drawn from the same test suspension and with the same acceptance criteria as previously.

6.3 Ruggedness

Although this data should be available from the equipment supplier who will have access to multiple models of the same equipment a limited ruggedness could be performed again using type cultures but varying the lots of any reagents used to ensure that the identifications obtained were independent of such changes. It may also be possible in a large organization where a method has been established at several sites to organise a "round robin" of testing using a consistent culture, but varying other parameters.

6.4 Robustness

Again this should be part of the portfolio supplied by the manufacturers but a limited in-house determination could be done by, for example, varying the age of the cultures when tested; age of any media or reagents used; using media at the extremes of the acceptable pH range etc.

With all of the above the main criteria being evaluated is equivalency (i.e., the proposed method is as consistent as the registered or pharmacopoeial method, when applied to well-characterized microorganisms in achieving an unequivocal identification). In our experience even the pharmacopoeial methods may not provide consistent identification.

6.5 Qualitative Methods (Presence or Absence)

For qualitative or presence or absence methods, the validation criteria suggested are:

- Accuracy
- Precision
- Limit of detection
- Ruggedness
- Robustness

Standard methods for detection of microorganisms are in themselves variable, particularly when low levels are being enumerated, making critical comparison of standard versus rapid methods particularly problematic.

The first problem is the preparation of a homogenous suspension from which successive representative samples can be drawn. This can be influenced by the type of organism, incubation conditions, length of incubation, media, etc., all of which may cause problems such as clumping of bacteria, chain elongation, or the formation of "gummy" exudates.

Before a comparison is attempted, time should be spent on determining the most consistent method of preparing a homogenous suspension. This may require different techniques for different microorganisms. In our experience, preparing such suspensions for fungi is even more problematic and may require aggressive homogenization techniques.

For existing methods, where presence or absence is based on turbidity, problems include the ability of bacteria to multiply to such an extent, that detection of turbidity may well be a function of the process by which they are removed from the substrate of interest (e.g., a cream); or the way in which any interference from the substrate is overcome. Additional or totally different manipulations needed for a rapid method may well confuse the equivalency.

Accuracy

Bearing in mind some of the issues identified, accuracy can probably only be performed by preparing very low concentrations of the target organisms (1 to 5 cfu/test unit) and inoculating these into a number of containers of the chosen media sufficient to obtain, after incubation, both positive and negative results. Accuracy of the rapid method is then based on providing at least the same degree of recovery, i.e., a similar relative proportion of both positive and negative results.

Precision

Precision can be determined by repeating the exercise on different lots of the same product. Due to the critical nature of presence or absence detection methods, e.g., sterility tests, we suggest that the validation be split into two phases. The first phase would be a comprehensive comparison across a wide range of target organisms, both pharmacopoeial and recent isolates with a range of products (if applicable). The second phase would be an ongoing comparison for a prolonged period or until a pre-determined number of comparisons performed side by side during routine usage had been reached. It is difficult to set a time or number against this, but for sterility testing it is suggested that 12 months or 100 separate tests may be appropriate.

Specificity

In this case specificity can be determined either by demonstrating that the rapid method can detect growth in the presence of the test article over a wide range of organisms (suitable for turbidimetric methods), or by demonstrating that the method does not erroneously detect the presence of extraneous matter from the test article and generate a positive result.

Limit of Detection

As for specificity the limit of detection comparison is performed by preparing low level inocula (1 to 5 cfu per test article) and demonstrating that the rapid method is as capable as the conventional method. At such low levels, a number of the replicates should show negative growth. We suggest that a range of organisms should be used, including those from the pharmacopoeia, plus the normal isolates from the test articles under comparison. Considerable replication is required to make this meaningful; not less than 10 replicates need to be performed for each organism used.

Ruggedness

The supplier of the rapid method should be able to supply data on the impact of using different instruments, different analysts, etc. However, this does not preclude the necessity of performing some measure of ruggedness in-house. Once again, the ability to prepare uniform samples significantly impacts on the discriminatory value of the test. The key test variables, e.g., analysts, reagents, time or temperature (where ranges are quoted), could all be challenged to show that under normal conditions such operational variability does not impact on the test's ability to correctly identify the presence (or absence) of a range of microorganisms.

Robustness

As for ruggedness the impact of small but deliberate variations in the method parameters, and their subsequent impact on the comparability of the methods, is probably best left to the manufacturer. As a guide the following would be expected to have been covered:

- Instrument performance over time
- Effects of different mixing times or techniques
- Effects of different incubation time or temperature variations within accepted limits
- Effect of ambient temperatures
- Effect of variability of any dispensing devices used
- Effect of lot-to-lot variation of reagents
- Effect of using reagents at end of shelf life

Much of this data has not been routinely generated for most conventional methods.

Quantitative Methods

It is for these techniques that the vagaries of microbiology, particularly in homogenous test sample preparation, really begin to bite! In order to demonstrate equivalency with the conventional method, the following parameters need to be evaluated:

- Accuracy
- Precision
- Specificity
- Limit of quantification
- Linearity
- Range
- Ruggedness or robustness
- Equivalence

Accuracy

Before evaluating this parameter, it is worth reviewing the difficulties associated with determining the closeness of a test result to the true value in a microbiological context. Table 7.1 lists the number of replicates needed to claim a 90% probability that the results between two methods differ from between 10 to 100% for a range of bacterial conditions.

Table 7.1 Number of Replicate Determinations Needed for Enumeration of Bacteria at Different Concentrations

% Difference	1 cfu	10 cfu	30 cfu	50 cfu	100 cfu	300 cfu
10	1887	189	(63)	38	19	6
20	517	52	14	10	5	2
25	345	35	12	7	3	1
50	106	11	4	2	1	1
100	37	4	1	1	1	1

7 CASE STUDY

If a technician wished to claim with a 90% probability, that two methods used to enumerate a suspension containing 30 cfu with an acceptance criteria of no more than a 10% difference in results, he would need to perform 63 replicates! There is little regulatory guidance as to what constitutes agreement, although *Ph Eur 2002* (2.6.12) states that where a limit of 10^2 is given then results up to 5×10^2 may be considered compliant. *USP 24*/NF 19 (1231) provides data on plate-count enumeration, stating that the error on a count of 3 cfu on a plate from a 10^{-1} dilution is 58%.

To determine accuracy, cultures of a number of organisms are prepared providing a range of dilutions, i.e., 100; 75; 50; 25; 10% of the original suspension, counts in the order of 30 to 300 cfu over the dilution range. Assuming the lowest count seen is around 30 and with an acceptance criteria of no more than a 25% difference in results, then 12 replicates of each dilution using the conventional and rapid method could be used.

Statistical methods of comparing the two data sets such as students t-test or analysis of variance could be used.

7.1 Precision

Precision is expressed as either the standard deviation (SD) or relative standard deviation (RSD) of the method. For a microbiological enumeration method values within 0.5 log are considered precise. To compare the precision of one method relative to another, a suspension should be prepared at the upper end of the test capability, which is then serially diluted down to the lower end of the range. Between 2 to 5 of the dilutions should be compared with at least 10 replicates of each being performed. An SD or RSD in the region of 10 to 15% is considered acceptable. The variance of each method can be statistically compared using the F test enabling any significant difference between the precision of the two methods to be demonstrated.

7.2 Specificity

In this context, specificity is the ability of the method to isolate and enumerate a range of microorganisms. Compendial organisms for fertility testing are supplemented with the most prolific organisms isolated from the products or samples routinely screened. The rapid method should be equally proficient as the traditional at enumerating the range of microorganisms.

7.3 Limit of Quantification

For the traditional plate count method the limit of quantification is 1 cfu/ml. However, the highest count accuracy is obtained when there are from 30 to 300 organisms per plate. If membrane filtration is the method used, then the limit is 1 cfu from the filtered volume. For example, for a large volume parenteral solution this may require filtering of between 1 to 5 litres. However, this limit can also be determined by any pretreatment required to solubilize or extract the microorganism from the sample.

So if 10-g sample has to be diluted in 90 ml of diluent and 1 ml enumerated using pour plates, then the limit could only be reported as less than 10 cfu/g of sample. At best the limit of quantification should be such that very low levels of microorganisms are detected and enumerated, with the same frequency for the rapid method as for the conventional. One way to do this is to carefully prepare a suspension, to contain as close as possible to 1 organism per ml. Multiple replicates (from 25 to 50) are enumerated by the conventional or rapid method. Even in the most carefully prepared suspension actual results achieved will probably be in the range of 0 to 5 cfus. Both methods are expected to detect and enumerate microorganism in this range equally successfully.

7.4 Linearity

Linearity is the ability of the test method to provide results proportional to the concentration of microorganisms present in a sample across a given range. Using a combination of the compendial organisms and routine isolates, cultures are prepared and diluted across the useable range of the methods being compared. For plate count, this would equate to approximately 300 organisms per plate at the top of the range, and a single organism at the bottom. At least five replicate determinations at each concentration across the range for each organism is used for both methods. The easiest way to then compare the data is graphically, by plotting the results obtained against the dilutions used taking the dilution providing the highest result as 100%. Alternatively the data can be statistically manipulated using the correlation coefficient r^2 to measure linearity. A value for r^2 of 0.9 or better should be expected.

7.5 Range

The range is the interval between the upper and lower levels of microorganisms enumerated with precision, accuracy and linearity using the selected method.

7.6 Ruggedness

As for qualitative methods, this data is probably best obtained from the equipment manufacturers, who will have access to multiple instruments, multiple reagent lots, etc. It would normally be demonstrated across a range of microorganisms performing a number of replicates (5 to 10), while the parameters chosen are varied. For a user, this study will probably be limited to using different analysts, different lots of reagents, and different lots of media. The impact could be measured using the coefficient of variation which should be in the range of 10 to 15%.

7.7 Robustness

Robustness, like ruggedness, is the demonstration of the ability of the chosen method to cope with variations. However, for robustness the critical test parameters are deliberately varied. For example, the concentration of any reagents used may be varied by ± 20%; the incubation temperatures could be used at the top or bottom of the acceptable range; mixing or holding times can be varied, and so on. The objective is to demonstrate across a range of microorganisms and dilutions, that such deliberate variations do not impact on the method's ability to accurately determine the number of microorganisms present.

7.8 Equivalence

All these tests are largely measures of equivalency, in that for each one the outcome, however compared, is expected to be the same within the limitation of microbiological methods. However, they are all performed to some extent on "artificial" samples (e.g., type culture collection organisms, pure cultures, predominantly laboratory adapted strains). This is not normally the situation where either conventional or rapid methods are used. Here, multiple organisms will be found, many will be fastidious in their nutritive requirements or will be stressed or sublethally damaged by the environment from which they are being isolated. Equivalency looks at the comparability of the methods, when used side by side, analysing the routine samples. If the samples are process or purified water, then comparison of daily samples over a period of 28 days could be used to demonstrate equivalency. If they are product, then at least three different lots of each one sampled should be compared.

8 READY RECKONER

The ready reckoner allows the user to quickly determine which validation tests are needed for which rapid method type.

Validation Parameter	Quantitive Test	Qualitative Test
Accuracy	Yes	No
Precision	Yes	No
Specificity	Yes	No
Limit of Quantification	Yes	No
Limit of Detection	Yes	No
Linearity	Yes	No
Range	Yes	No
Robustness	Yes	Yes
Equivalence	Yes	Yes

REFERENCES

R. Baird, N. Hodges, S. Denyer (eds). *Handbook of Microbiological Quality Control.* Taylor and Francis (first edition published by Ellis Horwood Limited, 1985).

Denyer, S.P. and Ward, K.H. *Journal of Parenteral Science and Technology* 1983; 37: 156–158.

Evaluation, Validation and Implementation of New Microbiological Testing Methods. *PDA Journal of Pharmaceutical Science and Technology* Technical Report No. 33, May/June 2000.

FDA CDER Advisory Committee for Pharmaceutical Sciences May 2002. *Rapid Microbial Testing.*

Guidance for Industry: Analytical Procedures and Methods Validation. FDA Draft Guidance August 2000.

K.-O. Habermehl. *Rapid Methods and Automation in Microbiology and Immunology.* Springer-Verlag, 1985.

International Committee for Harmonisation (ICH) Q2B 1997. *Note for Guidance on Validation of Analytical Procedures; Methodology.*

Lundin, A. In *ATP Bioluminescence*, pp.11–30. Blackwell, Oxford, 1989.

Rapid Microbiological Monitoring Methods: The Status Quo. International Water Association, "The Blue Pages," July 2000.

Wills, K., Woods, H. et al. Satisfying Microbiological Concerns for Pharmaceutical Purified Waters Using a Validated Rapid Test Method. *Pharmacopoeial Forum* 24(1), Jan/Feb 1998.

Stanley, P.E. *Journal of Bioluminescence and Chemiluminescence* 1992; 7: 77–108.

Index